Genes & Signals

Genes & Signals

Mark Ptashne

Memorial Sloan-Kettering Cancer Center

Alexander Gann

Cold Spring Harbor Laboratory

Cold Spring Harbor Laboratory Press
Cold Spring Harbor, New York

Genes
& Signals

Printed in the United States of America

Developmental Editor Siân Curtis
Project Coordinator Joan Ebert
Production Editor Dorothy Brown
Desktop Editor Susan Schaefer
Interior Designer Denise Weiss
Cover Designer Ed Atkeson

Front Cover (paperback edition): "Ora Pro Nobis," oil on canvas (60 x 45 inches) by Hans Hofmann, 1964 (©Estate of Hans Hofmann/Licensed by VAGA, NY; courtesy Ameringer Howard Yohe). (Photo by Tom Powel Imaging.)

Library of Congress Cataloging-in-Publication Data

Ptashne, Mark.
Genes and signals / by Mark Ptashne and Alexander Gann.
p. cm.
Includes index.
ISBN 0-87969-631-1 (cloth) -- ISBN 0-87969-633-8 (pbk.)
1. Genetic regulation. I. Gann, Alexander, 1962– II. Title.
QH450 .P83 2002
572.8′65--dc21

2001053889

10 9 8 7 6 5 4 3

All Cold Spring Harbor Laboratory Press publications may be ordered directly from Cold Spring Harbor Laboratory Press, 500 Sunnyside Boulevard, Woodbury, New York 11797-2924. Phone: 1-800-843-4388 in Continental U.S. and Canada. All other locations: (516) 422-4100. FAX: (516) 422-4097. E-mail: cshpress@cshl.edu. For a complete catalog of all Cold Spring Harbor Laboratory Press publications, visit our World Wide Web Site http://www.cshlpress.com

For
François Jacob
and
L and L

Contents

Preface

THIS BOOK, WE HOPE, SERVES TWO FUNCTIONS. It introduces the reader to general principles of how genes are controlled—that is, how genes are expressed, or not, in response to environmental signals. The book also places the matter in a larger context: the molecular mechanisms underlying gene expression apply to many other biological regulatory processes as well.

We have attempted to write so that a basic introduction to molecular biology will suffice for the reader to follow the arguments. The student who can read a previous effort of one of us (*A Genetic Switch*) should be able to read this book as well.

We face the strain of deciding where details illuminate or obscure the main points. We have, in the text, selected a limited set of examples, and included a flavor of the background and experimental strategies. Footnotes giving more details are found at the end of each chapter. We have attempted neither a complete nor a historical survey. We have restricted references here to review articles and books that discuss many of the experimental results that we allude to. Some of these articles also express viewpoints at variance with ours. The Web Site www.GenesandSignals.org makes available all the figures from the book, as well as four lectures based on the book delivered by Ptashne at Rockefeller University in January 2002. The web site also provides references to matters discussed in the footnotes.

The book is written as an extended argument, and each chapter assumes that the reader has read the previous chapters, including the Introduction.

E-mail has made book writing both easier and harder. Easier, because one can, with a few strokes, ask a dozen people around the world whether such and such is true, or what they think about this and that. We have found such correspondents—some of whom we have never met—to be extraordinarily generous. But e-mail also makes the process harder because, given the ease with which one can do so, one feels obliged to ask; and sometimes, of say ten answers, five will be yes, and five will be no. So we have done our best to get things straight. We express our gratitude to all these correspondents. We explicitly thank the following who read all or part of the text or helped us with the figures: Sankar Adhya, Bobby Arora, Mary Baylies, Alcides Barberis, Richard Brennan, Jackie Bromberg, Martin Buck, Lew Cantley, Mike Carey, David Gross, Dale Dorsett, Michael Elowitz, Gary Felsenfeld, Michael Fried, Marc Gartenberg, Ed Giniger, Joe Goldstein, Jim Goodrich, Mark Groudine, Steve Hahn, Ann Hochschild, Michael Jaffe, Sandy Johnson, Horace Judson, Roger Kornberg, Dennis Lohr, Tom Maniatis, Peter Model, Noreen Murray, Ben Neel, Geoffrey North, Nipam Patel, Richard Reece, Danny Reinberg, Marilyn Resh, Bob Schleif, Susan Smoleski, Dimitri Thanos, Becky Ward, Adam Wilkins, Michael Yaffee, Michael Yarmolinsky, Pengbo Zhou, Jianing Xia, members of the Ptashne lab, and, especially, Pete Broad, Amy Caudy, Richard Ebright, Monique Floer, Grace Gill, Alan Hinnebusch, Guillaume Lettre, Julie Reimann, and Bill Tansey. We thank Renate Hellmiss for the illustrations, Siân Curtis and John Inglis for helpful editing, and Tony Pawson for some early, particularly stimulating, discussions and for contributing the foreword. Also James Doon and Mary Jo Wright for their help, and Bertil Daneholt for urging us to undertake this project.

Mark Ptashne
Ludwig Professor of Molecular Biology
Memorial Sloan-Kettering Cancer Center
and Cornell University
New York, New York

Alexander Gann
Cold Spring Harbor Laboratory Press
and The Watson School of Biological Sciences
Cold Spring Harbor Laboratory
Cold Spring Harbor, New York

September, 2001

Foreword

AN ORGANISM IS A COMPLEX ASSEMBLY of different kinds of cells that perform many different functions. A major goal of biological research is to understand how that complexity is generated. The problem becomes especially fascinating when one considers that the human genome has surprisingly few genes, only a fewfold more than much simpler creatures. Where is all the extra information that accounts for this complexity? And as complex systems evolve, does each advance in cellular function require the invention of an entirely new way of doing things, or are pre-existing molecular devices re-used in more complex ways, much as standard bricks can be used to make increasingly palatial buildings?

In this book, Mark Ptashne and Alex Gann have addressed these key questions by focusing on the mechanisms through which genes are transcribed into RNA, and the regulatory steps that govern the complement of genes expressed in any given cell. If basic principles of cellular organization exist, they should be central to this most critical of all cellular activities. Ptashne and Gann build their case from bacterial systems, where various alternative ways of regulating gene expression can be found and experimentally distinguished. They then move on to more complicated nucleated cells, and argue that, with bacterial models in hand, it is possible to discern beneath all the extraordinary complexity of gene expression a process of rather simple molecular interactions that regulate transcription in these systems. The additional complexities typical of eukaryotes can be explained as "add-ons" that allow this simple mechanism to accommodate the greater regulatory demands of organisms like yeast and metazoans.

The authors then point out that imposing control and specificity through simple protein-protein interactions—what they call "adhesive interactions"—is a common feature of other regulatory systems in the cell.

For example, the principles they annunciate for gene expression apply equally well to signal transduction. This premise suggests that to evolve increasingly complex biological systems, it may not be necessary to invent many new kinds of gene products. Rather, more sophisticated functions can be achieved by, for example, increasing the number of interactions that any one protein can make, through the reiterated use of simple binding domains, thereby expanding the possibilities for combinatorial association.

This book develops a persuasive and cogent argument that the regulated recruitment of proteins into specific complexes, and the resulting targeting of the transcriptional apparatus to selected regulatory elements, is the fundamental mechanism for controlling gene expression. When viewed in the larger context of signal transduction, it seems evident that these kinds of protein interactions provide a principal means of controlling cellular function.

This book opens up the basic molecular language that cells use for their internal organization and to communicate with the outside world. This is important, and fascinating, for anyone interested in how cells work and how regulatory systems evolve.

October, 2001

Tony Pawson
Samuel Lunenfeld Research Institute, Mt. Sinai Hospital
Toronto, Canada

Introduction

ENZYMES CATALYZE THE CHEMICAL REACTIONS in cells: they drive metabolism, replicate DNA, read genes, convey signals from the outside of the cell to their destinations within, and so on. Life depends on the specificity of enzymes.

Each enzyme typically catalyzes just one reaction. β-galactosidase, for example, cleaves the sugar lactose, and the protein kinase Src adds a phosphate to tyrosine residues within certain proteins. Each of these enzymes performs only its own reaction—a kinase cannot cleave a sugar, for example.

Many enzymes perform their characteristic function upon just one substrate. For example, only lactose has the proper size and shape to be recognized by β-galactosidase. Even closely related sugars do not fit properly into the enzyme's active site, the part of the protein that performs the cleavage. But enzymes such as the kinase face a further specificity problem: there are many tyrosine residues on many different proteins, and the enzyme must choose which of these it should modify.

A general solution to this problem has become apparent in recent years: "recruitment." The idea is that adhesive interactions between the enzyme and the chosen substrate bring the two proteins together. The active site on the enzyme then works on the appropriate tyrosine residue(s) within that protein. The residues involved in recruitment are usually well separated from those that make up the enzyme's active site and from those in the target protein that are to be modified. For many enzymes, specificity depends on proper recruitment.

RNA polymerase faces the specificity problem in a way that is particularly striking. This enzyme transcribes (reads) genes, copying the information in the DNA into RNA. The enzyme can work on a wide array of genes (each a potential substrate), and the choice of substrate is *regulated*. In the bacterium *Escherichia coli*, for example, the enzyme transcribes some 3000 genes: certain genes are transcribed only at certain times, often as determined by signals received by the cell from its surroundings. In eukaryotes, similar RNA polymerases face the same regulatory task but on a larger scale. This example of substrate choice is called *regulation of gene expression* or, more casually, *gene regulation*.

How is specificity achieved and regulated in this case? How are genes regulated? We suggest that the clearest way to answer this question is to examine the mechanism of action of a class of proteins called transcriptional activators. These proteins bind to specific sites on DNA and "turn on" (activate) expression of the nearby gene. There are several classes of activators, as we will see. But members of the predominant class work in a simple, even crude, fashion: they are adaptors that recruit the enzyme to the gene. As we will explain, these activators impose specificity on polymerase by presenting two adhesive surfaces, one that binds DNA, and the other that binds the enzyme. Other regulators, called repressors, work in a variety of ways, the simplest of which is to block the binding of polymerase to the gene.

This picture resonates with descriptions of how specificity is imposed on many other enzymes. These include, in addition to the kinases mentioned above: phosphatases, which remove phosphates from proteins; ubiquitylating enzymes, which target proteins for destruction; splicing enzymes, which work on RNA destined to become messenger RNA (mRNA); various DNA polymerases; and proteases that trigger cell death. In many of these cases, the enzyme has alternative substrates; the choice of substrate is determined by recruitment, and that regulation lies at the heart of many important biological processes.

There are dangers in using these kinds of binding interactions to help impose specificity on enzymes. The system can, in principle, rather easily go awry—changes in concentration can, for example, erase the role of these specificity determinants, allowing the enzymes to work on inappropriate substrates. And so a seemingly bewildering array of "add-ons" has accumulated to ensure that these enzymes work as they should. When viewed in this way, we believe that although we cannot predict the complexities to be found in any given case, we can better appreciate the *nature* of the complexities that confront us.

Given the dangers and complexities, why have these enzymes evolved as they have? Are there more elegant, more secure ways to accomplish the same ends? We do not know with any confidence the answer to these questions. But one point is clear: it is not difficult to see how enzymes of the sort we describe here can acquire new specificities—how a kinase can be changed to work on a protein it did not recognize before, or how a polymerase can be directed to a gene it might not have read before. All that is required are changes in the adhesive surfaces of regulatory proteins, or of the corresponding surfaces attached directly to the enzyme. Put another way, the system is highly "evolvable." There is more to the story, of course, but perhaps that formulation gives a flavor of the general idea.

Two fundamental aspects of biology—development and evolution—illustrate the importance of gene regulation.

GENE REGULATION AND DEVELOPMENT

A human is made of cells with widely differing characteristics—compare, for example, muscle, blood, and neural cells. These characteristics are specified by genes, and yet each cell (with few exceptions) contains the same set of genes. Two familiar ideas explain this apparent contradiction. First, it is the products of genes—enzymes and structural proteins, for example—not the genes themselves that determine cell architecture and behavior. Second, expression of genes, the process that yields these products, is regulated, as we have already pointed out—not all genes are expressed in all cells all the time.

Genes expressed during formation of one part of the organism are often expressed during formation of other parts as well. What distinguishes hands and feet, for example, is not solely (or even largely) the expression of different genes, but rather expression of common genes, but at different times, in different places, and in different combinations.

Very often the choice of which genes are expressed in a given cell depends on signals received from its environment. For example, the protein called Sonic Hedgehog, secreted by one set of cells during development, induces the neighboring cells to express genes that result in the formation of motor neurons. Signals, like genes, are reused during development: for example, Sonic Hedgehog, secreted by another set of cells, induces *its* neighbors to express genes required for proper formation of limbs.

Many genes respond to combinations of signals: one signal that turns a gene on, for example, may be overridden by another signal that turns it off. And, in more complex organisms in particular, a gene might be turned on (or off) only if multiple signals converge, a matter referred to as "signal integration."

GENE REGULATION AND EVOLUTION

Changes in gene regulation can contribute dramatically to morphological diversity. For example, the most striking morphological difference between the plant teosinte and maize, its domesticated form, is accounted for by the change in expression pattern of a single gene. To take another example, the disparate uses of a forelimb in two species of Crustacea—feeding in one case and locomotion in the other—is determined by a difference in the expression pattern of a single gene. And although at the moment we cannot be precise about this, it is generally believed that mammals—humans and mice, for example—contain to a large extent the same genes; it is the differences in how these genes are expressed that account for the distinctive features of the animals.

As of some 500 million years ago, representatives of most animal phyla had appeared, including chordates, arthropods, and echinoderms. At that time, these phyla underwent a dramatic period of diversification called the Cambrian explosion, and enormous phenotypic changes have also occurred since then. According to one popular view, however, changes in patterns of gene expression (rather than evolution of new genes) have had an important, perhaps even determinative, role in generating much of that diversity. The idea that changes in gene regulation have important roles in evolution was formulated more than 20 years ago by, among others, Mary-Claire King and Alan Wilson and by François Jacob. These ideas have been extended and supported by the work of many evolutionary and developmental biologists since then.

The following surmise, offered with more evocative than literal intent, summarizes these matters: a relatively small number of genes and signals have generated an astounding panoply of organisms. Thus, the regulatory machinery must be such that it readily throws up variations—new patterns of gene expression—for selection to work on. And these variants must be produced without destroying what has already been selected. In other words, the underlying mechanisms of gene regulation must be highly "evolvable."

GENE EXPRESSION AND ITS REGULATION

Two steps of gene expression are essentially the same in all organisms: the typical gene is transcribed into messenger RNA and that mRNA is then translated into protein. All cells contain at least one form of RNA polymerase—the enzyme that transcribes DNA into RNA—and the machinery that translates the mRNA into protein. There are additional steps in the process of gene expression found in some, particularly more complex, organisms. For example, for many genes, the RNA transcript must be "spliced," a process that removes unwanted sequences; it must be chemically modified at one or both of its ends; and it must be transported out of the nucleus for translation. We know of examples of regulation that affect each of these steps and additional steps besides.

But the most pervasive form of gene regulation—from bacteria to higher eukaryotes—involves the initiation of transcription by RNA polymerase. Regulation of that step, as we mentioned earlier, is the main subject of this book: we are concerned with mechanisms that ensure transcription will be initiated at one rather than another gene under a specified set of conditions. The first critical insights were reported some 40 years ago, as we now recount.

The modern study of gene regulation was initiated in the 1950s by François Jacob, Jacques Monod, Andre Lwoff, and many coworkers, at the Institute Pasteur in Paris. They recognized that bacteria, like higher organisms, regulate expression of their genes. Thus, for example, although the bacterium *E. coli* does not undergo a developmental process analogous to that of higher organisms, it nevertheless does not express all of its genes all of the time.

Jacob and colleagues studied extensively two examples. The first involves the ability of *E. coli* to grow on a wide array of different sugars including lactose; and the second concerns the lifestyle of a bacterial virus (a bacteriophage) called λ. These two apparently unrelated cases have something in common: in each, genes that previously were silent (turned off) are expressed (turned on or activated) in response to specific environmental signals.

The gene encoding the enzyme β-galactosidase is silent until its substrate—lactose—is added to the medium; the gene is then transcribed (turned on) and the enzyme is synthesized. In the λ example, a set of about 50 phage genes (the so-called lytic genes) can be maintained in a bacterium in a dormant (unexpressed) state called lysogeny. Those genes are transcribed in response to ultraviolet (UV) irradiation.

A key insight of the Paris group was that gene regulation—the decision as to which gene to express—can be separated from the process of gene expression per se. They isolated mutants of the bacterium and the phage that were specifically deficient in the regulatory process. In so doing, they identified a class of genes—regulatory genes—whose products have as their sole function the regulation of expression of other genes. Thus, in those early experiments, they isolated mutants of *E. coli* that expressed the *lacZ* gene—which encodes β-galactosidase—whether or not lactose is present in the medium. In these mutants, *lacZ* expression has become "constitutive": the gene is expressed independently of the environmental signal that ordinarily is required to switch it on. Analogous mutants of λ express the phage lytic genes constitutively.

These constitutive mutants, in both the bacterial and phage cases, were of two classes. One class defined the regulatory genes whose products (called the Lac and λ repressors, respectively) keep the target genes turned off when the inducing signal is absent. The second class defined DNA sites near the target genes, called operators, where the repressors act. In the late 1960s and early 1970s, it was shown explicitly that the Lac and λ repressors are proteins that bind specifically to their respective operators on DNA. Operators typically overlap the sequence recognized by polymerase, called the promoter. Repressor bound at the operator excludes binding of the polymerase and hence prevents transcription of the gene.

ALLOSTERY

How are the actions of those repressors, and other regulatory proteins, controlled by extracellular signals? The general answer is that regulatory proteins undergo crucial changes in shape in response to signals. For example, the Lac repressor bears, in addition to its DNA-binding surface, a pocket that binds allolactose (a metabolic derivative of lactose). Binding of allolactose exerts an effect on the conformation (shape) of the repressor hunter.bkstore.com

217808 CASH [illegible] 0043 [illegible]

9780879696337 NEW
PTASHNE/GENES&SIGNALS [illegible] [illegible]
TOTAL [illegible]

ACCOUNT NUMBER XXXXXXXXXXXX[illegible]20
Visa/Mastercard [illegible]
Expiration Date XX/XX
Authorization [illegible]

TEXTBOOKS RETURNABLE FOR 3 DAYS

9/16/08 [illegible] AM

such that, in the absence of lactose, the repressor binds DNA, but in its presence does not. The site on the repressor that binds the sugar is distinct from that which binds DNA; and because the protein changes shape, the phenomenon was called *allostery* (other shape). The site to which (in this case) the sugar binds was called the *allosteric site.*

Allostery is a profoundly important mechanism by which signals can be interpreted by organisms. Some enzymes, for example, are also controlled by allostery. Because the allosteric site need not, and typically does not, resemble the active site, any signal can, in principle, be used to control the function of any appropriately designed protein.

GENE ACTIVATION

The early description of the *lac* and λ systems was incomplete in one important way: the *lac* genes, and certain λ genes, are subject to activation as well as to repression. That is, even in the absence of the repressors, these genes are not fully expressed unless they are activated. Specific DNA-binding proteins effect this "positive control" just as specific binding proteins effect repression. As with repressors, the typical activator senses a signal that determines whether or not it binds DNA and activates transcription.

For reasons that we hope will become clear, a coherent view of gene regulation requires that we first understand molecular mechanisms of gene activation.

OVERVIEW OF THE BOOK

In Chapter 1 we describe three different mechanisms found in *E. coli* for activating genes. We described one of these at the beginning of this Introduction, and it is disarmingly straightforward: the activator recruits RNA polymerase to a specific gene where the transcription reaction—a complicated affair—then proceeds spontaneously. "Regulated recruitment," as we call this mechanism, readily lends itself to the use of repressors, together with activators, to control specific genes, and to the use of regulators in different combinations (combinatorial control). Bacteria provide us with two additional mechanisms for gene activation and with experimental strategies for distinguishing between all three.

In Chapter 2 we apply those strategies to analyzing gene regulation in yeast, a eukaryote that is particularly easy to work with. We find that regulated recruitment is an important, perhaps predominant, mechanism for gene activation. In this case, in addition to RNA polymerase, other enzymes are recruited—enzymes that modify chromatin, for example. This provides opportunities for signal integration and combinatorial control not found in bacteria.

In Chapter 3—a very brief survey of higher eukaryotes—we see even more extensive combinatorial control and signal integration, and despite many unresolved issues, at the heart of the matter lies regulated recruitment. Here, as in the previous two chapters, we will see a variety of means by which signals are transmitted from the environment to control the activities of activators and repressors.

In Chapter 4 we consider how recruitment is used to impose specificity on some of the other enzymes we mentioned earlier—kinases, for example. We discuss some of the general consequences of using regulated recruitment to impose specificity on disparate enzymes.

A note on nomenclature: we follow commonly accepted convention according to which the name of a gene is italicized but its protein product is not. There are idiosyncrasies, however. For example, in bacteria the gene is in lowercase but in yeast it is in uppercase.

BIBLIOGRAPHY

Books

Beckwith J.R. and Zipser D., eds. 1970. *The lactose operon*. Cold Spring Harbor Laboratory, Cold Spring Harbor, New York.

Bier E. 2000. *The coiled spring: How life begins*. Cold Spring Harbor Laboratory Press, Cold Spring Harbor, New York.

Cairns J., Stent G.S., and Watson J.D., eds. 1992. *Phage and the origins of molecular biology* (Expanded edition). Cold Spring Harbor Laboratory Press, Cold Spring Harbor, New York.

Carroll S.B., Grenier J.K., and Weatherbee S.D. 2001. *From DNA to diversity: Molecular genetics and the evolution of animal design*. Blackwell Science, Malden, Massachusetts.

Davidson E.H. 2001. *Genomic regulatory systems: Development and evolution*. Academic Press, San Diego, California.

Gerhart J. and Kirschner M.W. 1997. *Cells, embryos, and evolution: Toward a cellular and developmental understanding of phenotypic variation and evolutionary adaptability*. Blackwell Science, Malden, Massachusetts.

Gilbert S. 1997. *Developmental biology*, 5th edition. Sinauer Associates, Sunderland, Massachusetts.

Hershey A.D., ed. 1971. *The bacteriophage lambda*. Cold Spring Harbor Laboratory, Cold Spring Harbor, New York.

Judson H.F. 1996. *The eighth day of creation: Makers of the revolution in biology* (Expanded edition). Cold Spring Harbor Laboratory Press, Cold Spring Harbor, New York.

Lwoff A. and Ullmann A., eds. 1979. *Origins of molecular biology: A tribute to Jacques Monod*. Academic Press, New York.

Raff R.A. 1996. *The shape of life: Genes, development, and the evolution of animal form*. University of Chicago Press, Chicago, Illinois.

Stahl F.W., ed. 2000. *We can sleep later: Alfred D. Hershey and the origins of molecular biology*. Cold Spring Harbor Laboratory Press, Cold Spring Harbor, New York.

Wolpert L. 1998. *Principles of development*. Oxford University Press, New York.

Review Articles

Browne W.E. and Patel N.H. 2000. Molecular genetics of crustacean feeding appendage development and diversification. *Semin. Cell Dev. Biol.* **11:** 427–435.

Carroll S.B. 2000. Endless forms: The evolution of gene regulation and morphological diversity. *Cell* **101:** 577–580.

Doebley J. and Lukens L. 1998. Transcriptional regulators and the evolution of plant form. *Plant Cell* **10:** 1075–1082.

Duboule D. and Wilkins A. 1998. The evolution of 'bricolage'. *Trends Genet.* **14:** 54–59.

Jacob F. 1977. Evolution and tinkering. *Science* **196:** 1161–1166.

Jacob F. and Monod J. 1961. Genetic regulatory mechanisms in the synthesis of proteins. *J. Mol. Biol.* **3:** 318–356.

Monod J. 1966. From enzymatic adaptation to allosteric transitions. *Science* **154:** 475–483.

Tautz D. 2000. Evolution of transcriptional regulation. *Curr. Opin. Genet. Dev.* **10:** 575–579.

CHAPTER 1

Lessons from Bacteria

[He], like the rest of us, had many impressions which saved him the trouble of distinct ideas.

GEORGE ELIOT

Classical models tell us more than we at first can know.

KARL POPPER

WE BEGIN WITH A BRIEF DESCRIPTION of the enzyme RNA polymerase and a summary of the three mechanisms of gene activation found in *Escherichia coli*. We then describe the mechanisms in more detail, using examples of each, and include a description of the role of repressors where appropriate. We will also see, for each case, how signals are transmitted from environment to gene.

We pay particular attention to the nature of the molecular interactions required in each case. We will encounter experimental approaches that distinguish between the mechanisms. As shown in subsequent chapters, several of these tests can be applied to analyzing gene regulation in eukaryotes as well.

RNA POLYMERASE

Four subunits comprise the core of RNA polymerase (see Figure 1.1). When examined in vitro, this complex transcribes DNA into RNA, but initiates at nonspecific sites on the DNA.

In the bacterial cell, the core is typically found to be associated with one other essential subunit, called Sigma (σ), and the complex is referred to as the holoenzyme. σ imposes a level of specificity: it restricts initiation of transcription to promoter sequences. There are six or seven different σ

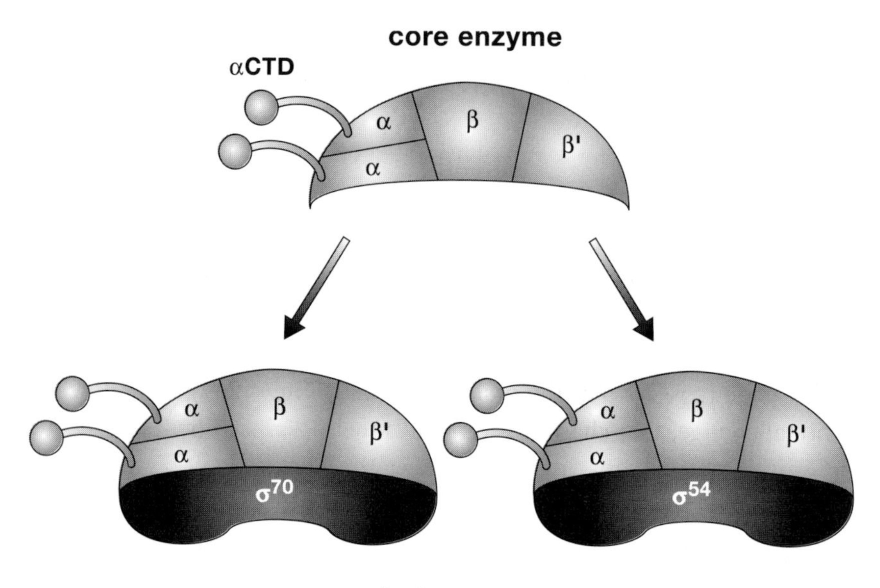

FIGURE 1.1. The *E. coli* RNA polymerase. Two of the four core subunits (β and β′) are present in single copy, and the other (α) is present in two copies. Their names reflect their sizes—σ^{70} is some 70 kD. α has two domains—the smaller carboxy-terminal domain (CTD) is attached by a flexible linker to the amino domain, which is embedded in the main body of the enzyme. The enzyme's active site is found at the interface of the β and β′ subunits. In cells, the enzyme is attached to one or another σ subunit.

subunits, each of which directs the enzyme to a unique set of promoters. Holoenzyme bearing one σ (σ^{70}) is the predominant form and is responsible for transcribing the majority of the genes.[1]

There is then a further specificity problem. Which of the promoters any given holoenzyme *can* use *is* chosen under any given circumstance? Here, in outline, are three ways in which the choice is made.

- ***Regulated Recruitment:*** RNA polymerase bearing the σ^{70} subunit is constitutively active; activators work by recruiting that enzyme to specific genes. These activators often work in conjunction with repressors that, in their simplest mode of action, exclude binding of polymerase to specific genes. The genes we will consider include those required for metabolism of the sugar lactose and others involved in the growth of the bacterial virus λ.

- ***Polymerase Activation:*** RNA polymerase bearing the σ^{54} subunit binds to specific genes in a stable but inactive state. Activation of a particular gene requires that the enzyme attached to that gene literally be "activated," an effect that requires an allosteric change in that enzyme: the activator works by inducing that change. The genes we will consider are involved in nitrogen metabolism.
- ***Promoter Activation:*** RNA polymerase bearing σ^{70} (the same form as that which transcribes the *lac* and λ genes) also binds a small set of genes at which it forms a stable, inactive complex. Activation of a specific gene requires that the *promoter* of that gene undergo a conformational change; that change in the promoter is induced by the activator. Genes in this category encode proteins that render the cell resistant to one or another poisonous metal (e.g., mercury).

As we shall see, for regulated recruitment, simple "adhesive" interactions of the activator with DNA and with polymerase suffice, whereas in the other two cases, more specialized kinds of interactions are required for gene activation.

REGULATED RECRUITMENT: THE *lac* GENES

The enzyme β-galactosidase, the product of the *lacZ* gene, cleaves lactose, the first step in metabolism of that sugar. The gene is transcribed if, and only if, lactose is present in the medium. But that physiological signal is almost entirely overridden by the simultaneous presence of glucose, a more efficient energy source than lactose. Only after having exhausted the supply of glucose does the bacterium fully turn on expression of *lacZ*.

The effects of these two signals (lactose and glucose) are mediated by two DNA-binding regulatory proteins, each of which senses one of the sugars. Lac repressor binds the operator only in the absence of lactose, and CAP, an activator, binds DNA only in the absence of glucose.

The arrangement of regulator binding sites in front of the *lacZ* gene is shown in Figure 1.2. The states of the *lac* genes as found under different environmental conditions are shown in Figure 1.3 and are as follows.

1. In the presence of both sugars, neither regulatory protein binds DNA, and RNA polymerase transcribes the *lac* genes at a low level. That

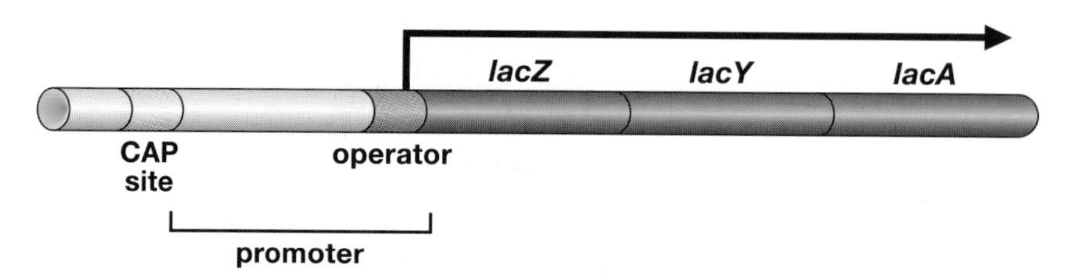

FIGURE 1.2. The *lac* operon. The *lacZ* gene is transcribed in a single mRNA along with two other genes, *lacY* and *lacA*. *lacY* encodes the permease that brings lactose into the cell, and *lacA* encodes an acetylase that is believed to detoxify thiogalactosides, which, along with lactose, are transported into cells by *lacY*. The promoter spans about 60 bp, and the CAP site and the operator (the Lac repressor-binding site) are about 20 bp each. The operator lies within the promoter, and the CAP site lies just upstream of the promoter. The picture is simplified in that there are two additional, weaker, *lac* operators located nearby (see Repression by *Lac* Repressor). It is not to scale: the *lacZ* gene, e.g., is about 3500 bp long. The entire element shown is called an "operon."

level—called the basal level—is determined by the frequency with which the enzyme spontaneously binds the promoter and initiates transcription.

2. In the presence of lactose and the absence of glucose, CAP is bound to its DNA site, but repressor is not. Transcription of the genes is activated: that is, they are transcribed at a level some 40-fold higher than the basal level. CAP has this effect by recruiting RNA polymerase to the promoter. Polymerase binding and transcription initiation are the same as in the basal case—they simply happen more frequently.

3. In the absence of lactose, repressor is bound to the operator, polymerase is excluded from the promoter, and transcription is essentially abolished. This is true whether or not glucose is present, and therefore whether or not CAP is active.

In sum, we have a constitutively active enzyme (RNA polymerase) that, alone, works with a certain frequency. The activator increases this frequency by recruiting the enzyme to the gene, and the repressor decreases the frequency by excluding the enzyme.

We now turn to some of the molecular details of these reactions and describe experiments that support the description we have given.

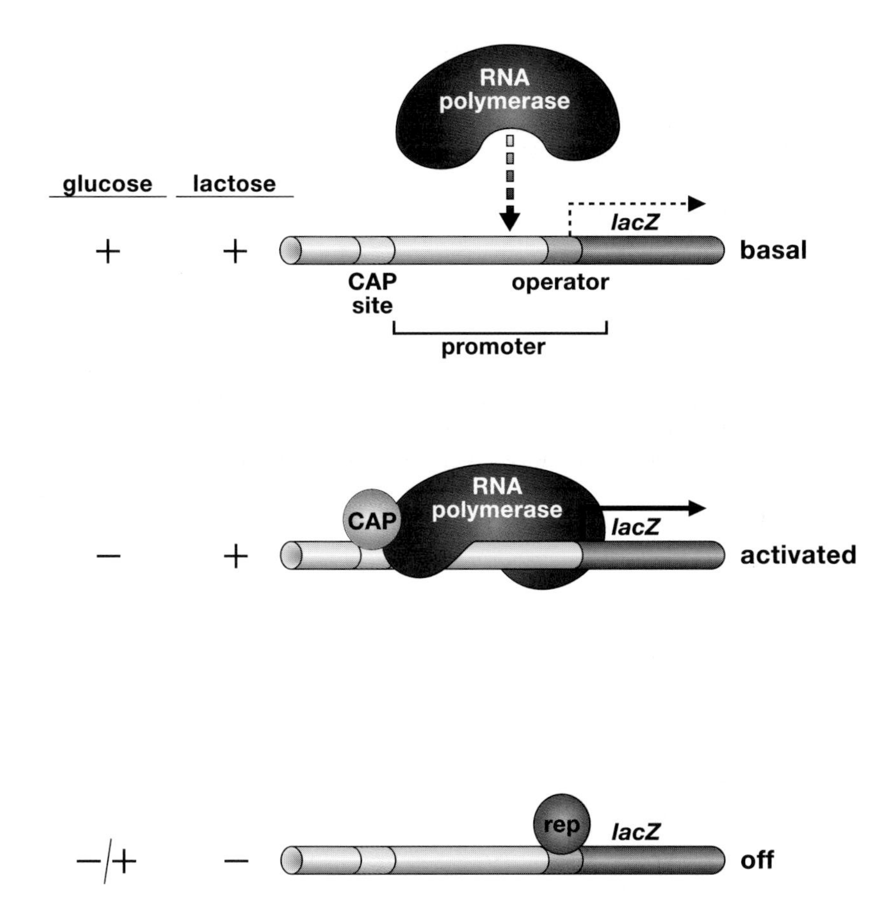

FIGURE 1.3. Three states of the *lac* genes. When bound to the operator, repressor excludes polymerase whether or not active CAP is present. Polymerase is shown in a more simplified form than in Figure 1.1: the carboxyl domains of the α subunits, which protrude from the left end of the polymerase as drawn here, are not shown. CAP actually binds DNA as a dimer, and Lac repressor as a tetramer.

Protein-DNA Interactions

Both CAP and Lac repressor dock with their DNA sites using a similar structural motif, the so-called helix-turn-helix (HTH). The schematic representation of Figure 1.4 shows a protein dimer bound to DNA. Each HTH bears one α-helix (the "recognition helix") that inserts into the major groove of DNA. The side chains of amino acids exposed along the

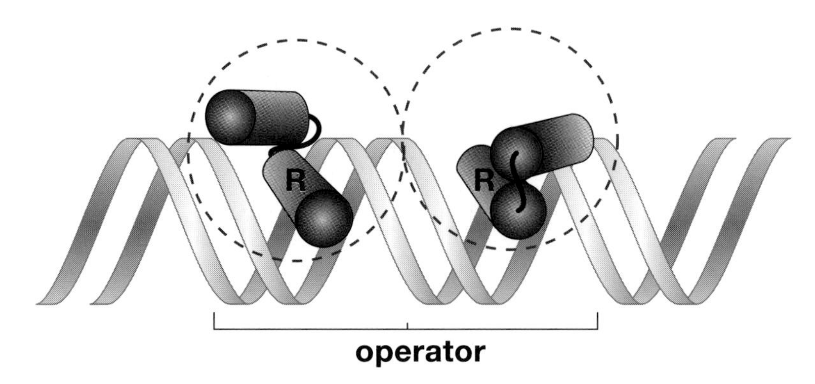

FIGURE 1.4. DNA binding by a dimeric protein bearing a helix-turn-helix domain. The dotted circles represent two identical subunits of a DNA-binding protein complexed with an operator. The HTH motif on each monomer is indicated, with the "recognition helix" labeled R. A single recognition helix can contact functional groups displayed on the edges of approximately 6–8 bp. A single simultaneous use of two HTH motifs renders the binding far more specific than were a single HTH to be used. (Modified, with permission, from Ptashne 1992.)

recognition helix make sequence-specific contacts with edges of base pairs. A second helix lies across the DNA; it helps position the recognition helix and strengthens the binding. Differences in the residues along the outside of recognition helices largely account for differences in the DNA-binding specificities of regulators.[2]

The HTH motif is the predominant DNA-recognition module found among *E. coli* transcriptional regulatory proteins. A somewhat modified form is found in eukaryotes in so-called homeodomain proteins.

Detecting Physiological Signals

As we noted in the Introduction, Lac repressor undergoes a conformational (allosteric) change upon binding inducer (allolactose). This change greatly decreases DNA binding of the repressor. In contrast, the allosteric change undergone by CAP upon binding a small molecule (cyclic AMP or cAMP) *increases* its ability to bind DNA. Glucose exerts its effect by (somehow) decreasing synthesis of cyclic AMP. The name CAP stands for catabolite activator protein; the same protein is often called CRP, for cyclic AMP receptor protein.[3]

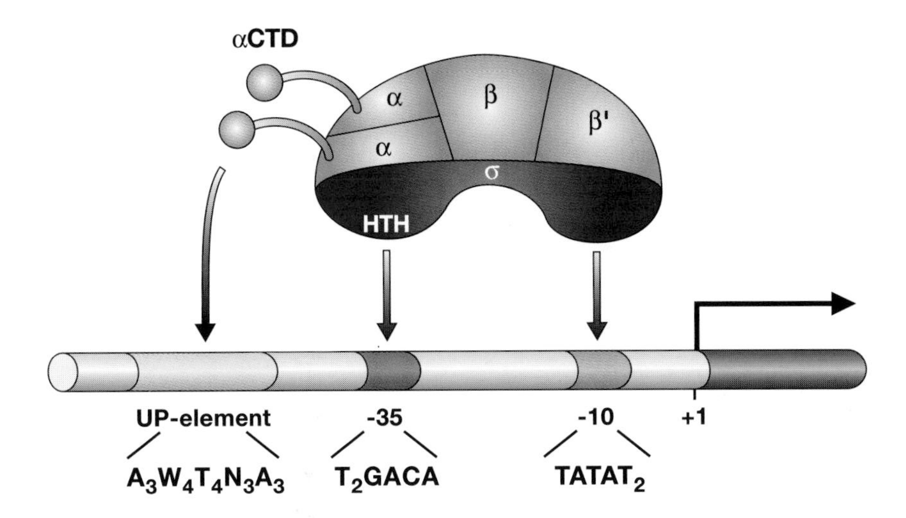

FIGURE 1.5. Sequence elements in promoters recognized by holoenzyme containing σ^{70}. The σ subunit is an elongated protein that can simultaneously contact the –10 and –35 regions of the promoter. Where present, the UP-element is recognized by the carboxy-terminal extensions of the α subunits. Although not indicated here, polymerase covers the transcription start site and an additional 20 bp downstream. Beneath each element is shown a "consensus" sequence for that element. W represents adenine or thymine; N represents any base; and a subscript indicates the number of reiterations of that base.

Promoter Recognition and Transcription by RNA Polymerase

The *lac* promoter contains two sequence elements that are recognized by the σ^{70} subunit of polymerase. The sequence in the –35 region (counting back from the transcription start site) is recognized by an HTH domain; the other sequence, in the region around –10, is recognized by a different part of σ.

All promoters recognized by a σ^{70}-containing holoenzyme bear a sequence related to the canonical –10 region shown in Figure 1.5. Some, like the *lac* promoter, also bear sequences related to the canonical –35 region. Some, unlike lac, also bear a so-called UP-element, positioned upstream of the –35 region, which is recognized by the carboxyl domain of the α-subunit.

All three elements are found at some promoters, and where both upstream elements are lacking, the –10 region is extended by one or two characteristic base pairs. In general, the more elements present at a promoter, and the more closely they resemble the canonical sequences, the more efficiently it works. Some promoters, as we will see, work at very high levels without the aid of an activator.

Polymerase initially binds to promoter DNA in its ordinary double-stranded (or "closed") state. That complex then undergoes a structural transition to the "open" form in which approximately 14 base pairs (bp) around the transcription start site are melted ("opened") to expose the template strand. Whereas formation of the closed complex is readily reversible, formation of the open complex is generally irreversible and, in the presence of the appropriate RNA precursors (nucleoside triphosphates), leads to transcription.

As we have discussed, despite all these complexities associated with transcriptional initiation, CAP activates transcription simply by recruiting the polymerase to the promoter. We turn now to the molecular interactions required for that recruitment.

Switching the Genes On: Activation by CAP

Polymerase recruitment, as shown in Figure 1.3, requires formation of a tripartite complex comprising CAP, polymerase, and DNA. The formation of that complex is an example of cooperative binding of proteins to DNA.

The term cooperative binding can have connotations other than those we intend here (see Appendix 1, More on Cooperativity), but the form of cooperativity we now describe has a crucial role in many aspects of gene regulation in many organisms.

Cooperative Binding of Proteins to DNA

In the example shown (Figure 1.6a), neither site A nor site B is efficiently occupied when protein A or B is present alone, i.e., were a series of snapshots taken of site A (in the presence of protein A alone), protein A might be found bound in, say, 1% of those pictures. This frequency would be increased were the concentration of protein A to be increased.

When both proteins are present, however, both sites are much more likely to be occupied even at the lower protein concentrations. This "cooperative" effect depends on the simultaneous interaction of each protein with its DNA site and with the other protein. We say the proteins cooper-

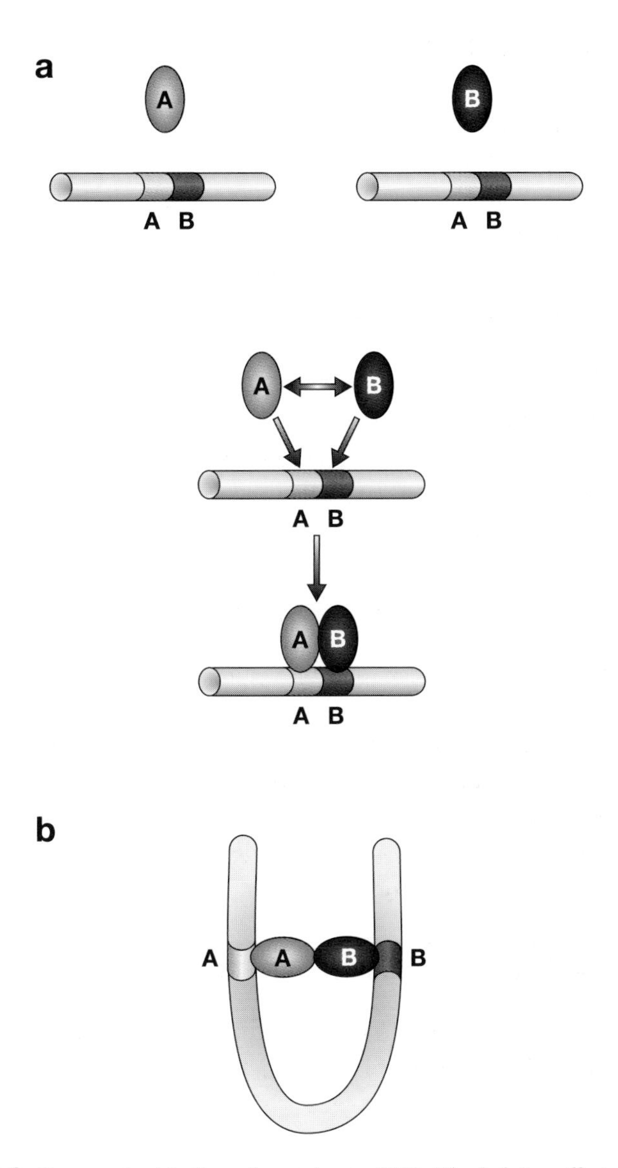

FIGURE 1.6. Cooperative binding of proteins to DNA. The helping effect requires that interaction between the proteins be compatible with their positions on DNA. Thus, for example, interacting proteins whose binding sites lie on, or nearly on, the same face of the helix will generally bind cooperatively more readily than if their sites are on opposite sides of the helix. This geometric restriction is eased if the DNA-binding and protein-protein interacting surfaces are carried on separate domains that are connected by flexible linkers.

ate, in that they help each other bind to DNA. The following are four important aspects of this reaction.

- The effect requires neither the utilization of energy (e.g., hydrolysis of ATP) nor conformational change in either protein; the proteins need only "touch" one another. To understand this, consider the frequency with which each protein falls off the DNA when the proteins are bound cooperatively. One protein might dissociate from the DNA, but its continued interaction with the other, DNA-bound, protein would hold it nearby, ensuring that it readily rebinds to its DNA site. For the protein to completely dissociate, it would have to let go of the DNA and the other protein simultaneously, or both proteins would have to simultaneously let go of the DNA. It is sometimes said that one protein helps another bind by increasing the latter's local concentration in the vicinity of its site.[4]
- The interaction between the proteins is weak in energetic terms (1 or 2 kcal). Interactions of this magnitude require very few specific contacts, but they can have important physiological consequences. A kilocalorie of interaction energy between proteins binding to separate DNA sites roughly corresponds, in terms of site occupancy, to increasing the concentration of either protein alone approximately 10-fold. And a change of 10–100-fold in binding can determine the difference between a site being essentially vacant or essentially fully occupied.
- For cooperative binding to be used as a control mechanism, the concentrations of the interacting partners must be held below a specified level—that at which they bind unaided—but high enough that they bind DNA in the presence of their respective partner. Precisely what these relevant concentrations are depends on the affinities of the proteins for their DNA sites, the strengths of the interactions between them, and so on.
- If the binding sites for A and B are not adjacent, cooperative binding would require that the intervening DNA "loop out" to accommodate the reaction as shown in Figure 1.6b.[5]

For further discussion of these matters, see Appendix 1, More on Cooperativity.

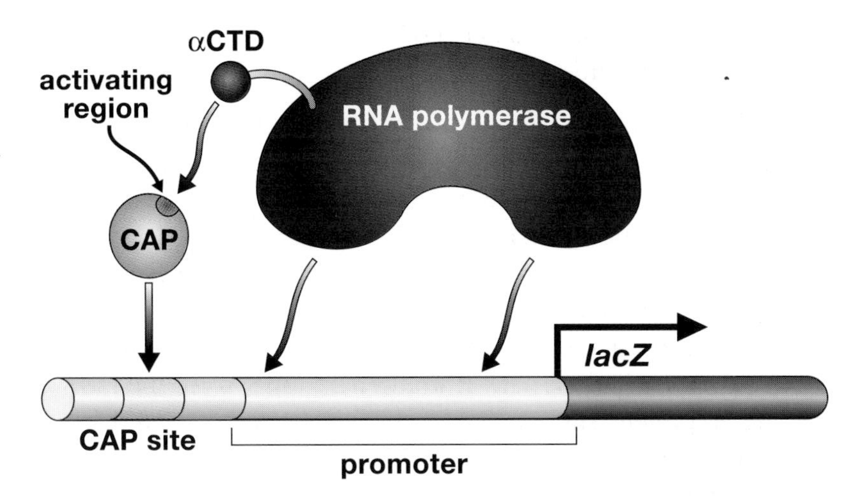

FIGURE 1.7. Cooperative binding of CAP and polymerase to DNA. The picture is simplified in that only one CAP monomer and only one αCTD are shown. It is also believed that in the absence of an UP-element (see Figure 1.5), αCTD is bound to DNA nonspecifically when contacted by CAP.

Cooperative Binding and Gene Activation by CAP

To apply the reactions depicted in Figure 1.6 to the case at hand (i.e., the stimulatory effect of CAP on transcription of the *lac* genes), we replace protein A with RNA polymerase and protein B with CAP (Figure 1.7). CAP recruits polymerase to the promoter by binding cooperatively with it. CAP holds polymerase there until the complex isomerizes to the open form. At that stage, the reaction becomes essentially irreversible, and initiation of transcription proceeds.

The physiological consequences of the CAP-polymerase interaction are dramatic—a 40-fold increase in expression—but the interaction is small in energetic terms. Indeed, it probably involves no more than a kilocalorie or so of interaction energy and has not been detected in the absence of DNA. Here are some of the key experimental findings that support these conclusions concerning the mechanism of action of CAP at the *lac* genes.

In vivo footprinting. A variety of techniques, lumped under the term "footprinting," can reveal whether DNA molecules bearing a binding site for a protein actually bind that protein at any given moment. In particular, footprinting can be used to assay polymerase occupancy of the *lac* promoter in a population of bacterial cells. It is found that, in the absence of functional CAP and Lac repressor, RNA polymerase can be detected at the *lac* promoter in only a very few cells, consistent with the low level of expression seen under those circumstances. But in the presence of functional CAP (and in the absence of repressor), virtually all of the cells bear polymerase at their *lac* promoters. Thus, CAP "recruits" polymerase to the promoter.

Positive control mutants. CAP contains two surfaces required for activation of the *lac* genes: one is the DNA-binding surface, and the other is its "activating region," which contacts RNA polymerase. The functions of these two surfaces can be distinguished by mutation. Mutants of CAP defective in their polymerase-binding surface, but which retain the ability to bind DNA, cannot activate transcription. These are called *pc* (for "positive control") mutants. Thus, activation is not solely a consequence of DNA binding by the activator (as it would be were the effect mediated through changes in local DNA structure). Rather, the *pc* mutants suggest that a protein-protein interaction between the DNA-bound CAP and polymerase is required for activation.[6]

Chemical cross-linking. Chemical cross-linking experiments reveal that the activating region of CAP, as defined by the *pc* mutants, interacts with the carboxyl domain of the α-subunit of polymerase. The cross-linking is observed when both CAP and polymerase are bound to DNA.

Polymerase mutants. Deletion of the carboxyl domain of the α-subunit of polymerase eliminates activation by CAP as assayed in vitro. Point mutations in the αCTD can also abolish activation by CAP.

DNA-binding mutants. Mutants of CAP that cannot bind DNA do not activate transcription. This is the result expected if CAP must recruit polymerase to DNA.

The idea that CAP recruits polymerase to the promoter—but has no additional effect—makes a further striking prediction: Alternative ways of bringing polymerase to the promoter should activate the gene. The following "activator bypass" experiments show that this is the case. Each of these, with the exception of the last, has been performed both in vivo and in vitro.

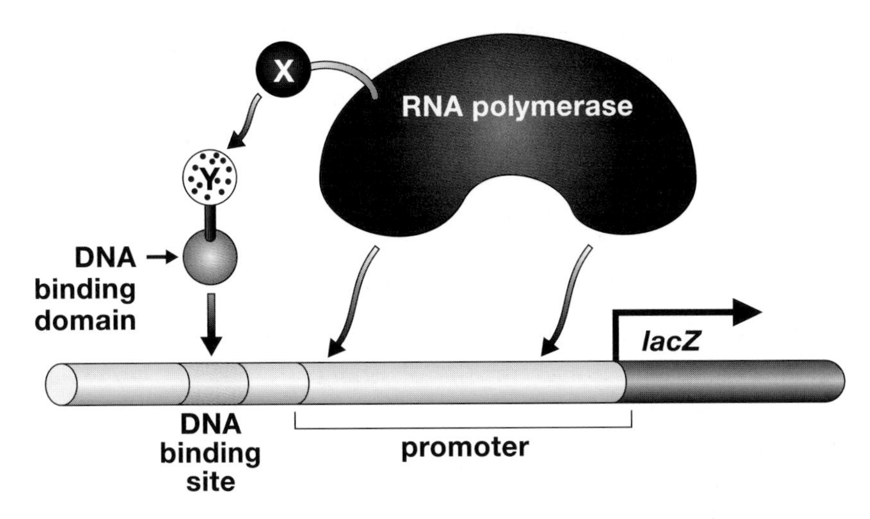

FIGURE 1.8. Activator bypass experiment 1. In this example, the αCTD has been replaced by another protein domain, labeled *X*, that is known to interact with protein *Y*. Protein *Y* has been fused to a DNA-binding domain, and the site recognized by that domain has been placed near the *lac* genes. Examples of such experiments include a case in which *X* is a domain of a yeast protein (called Gal11P), and *Y* is a domain of a yeast protein that interacts with Gal11P (the dimerization region of Gal4). These proteins are discussed in Chapter 2. In another case, both *X* and *Y* can be the interacting carboxyl domains of λ repressor, discussed later in this chapter.

- ***Activator bypass 1: Activation through heterologous protein:protein interactions.*** The CAP-polymerase interaction can be replaced by another protein-protein interaction. The experiment is performed as follows (Figure 1.8): one of two interacting proteins (*Y*) is attached to a DNA-binding domain, and the other interacting protein (*X*) is attached to RNA polymerase. Binding of the first hybrid protein to a site near the *lacZ* gene activates transcription in the presence of the modified polymerase. The key point is that the "heterologous" interaction (between *X* and *Y*) is not normally involved in transcriptional activation but nevertheless substitutes very well for the ordinary CAP-polymerase interaction in this experiment. Some examples of *X* and *Y* are given in the legend to Figure 1.8.
- ***Activator bypass 2: Direct tethering of polymerase.*** The protein-protein interaction between CAP and polymerase can be replaced by a protein-DNA interaction. In these experiments, a CAP DNA-binding domain is

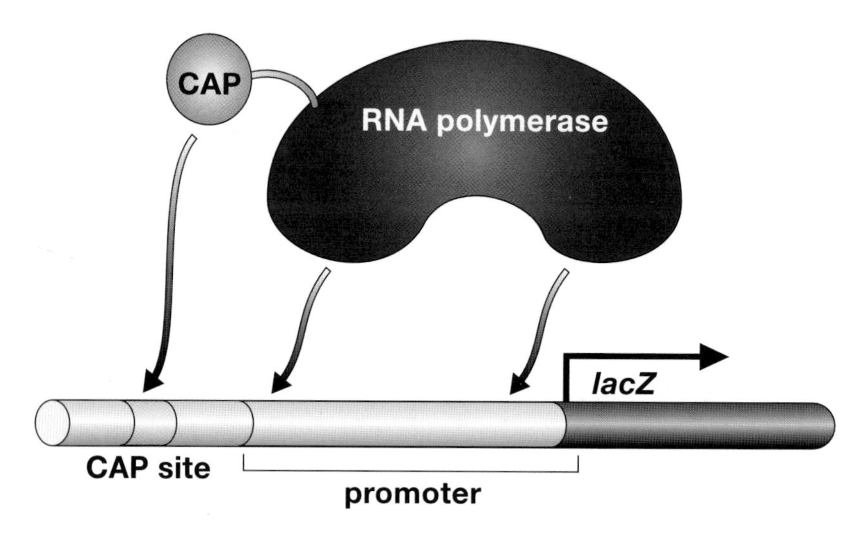

FIGURE 1.9. Activator bypass experiment 2. In this example, the αCTD has been replaced by the DNA-binding portion of CAP. Other experiments show that attaching a DNA-binding domain to ω (see Footnote 1) also works in such an experiment.

fused directly to RNA polymerase as shown in Figure 1.9, and high levels of transcription are achieved if the promoter bears a binding site for CAP. The activating region of CAP is not required for this activation. Thus, increasing the affinity of part of polymerase for a site near the promoter is sufficient to induce high levels of transcription.

Another way to introduce a new polymerase-DNA contact at the *lac* promoter is to change not the protein, but the DNA. As we have noted (see Figure 1.5), the carboxyl domain of the α-subunit is itself a DNA-binding domain: it binds to the UP-element found upstream of the –35 region in various strong promoters. Introducing this sequence upstream of the *lac* promoter renders efficient transcription independent of CAP.[7]

- ***Activator bypass 3: Increased concentration of polymerase.*** High levels of transcription in the absence of activator can also be obtained by increasing the concentration of polymerase, an experiment performed in vitro.

All of the experiments described here, taken together, are consistent with the idea that CAP works simply by recruiting the polymerase to the promoter. We cannot absolutely exclude the possibility that CAP also has effects on the structures of DNA and/or polymerase that contribute to activation, but such effects would have to be subtle.[8]

Repression by Lac Repressor

As we have already noted, the *lac* operator overlaps the *lac* promoter, and the repressor bound there blocks binding of polymerase to the promoter. In principle, any other protein bound equally tightly at this site would have a similar effect.[9]

Our description has been simplified by omitting the fact that there are actually two Lac repressor-binding sites in addition to the primary one we have described. These can contribute to repression, as we now explain.

The two additional *lac* operator sites are positioned about 90 bp upstream and 400 bp downstream from the primary operator. Deletion of either site decreases repression some 2–3-fold, and deletion of both reduces repression about 50-fold (i.e., from a factor of ~10,000 to a factor of ~200). A single Lac repressor—a stable tetramer—can simultaneously contact the primary operator and one or the other of the secondary operators, with the DNA looping out to accommodate the binding. Simultaneous binding to multiple operators strengthens the binding, and perhaps the looped structure itself contributes to repression (see Appendix 1, More on Cooperativity). These secondary operators, in the absence of the primary site, bind the repressor weakly and mediate virtually no repression.

Interim Summary and Extension

According to our formulation of the action of CAP, the term "gene activation" might be misleading in the following sense: the term "activation" might suggest to some that either the gene or the polymerase is being switched from an inactive to an active state—literally "activated." But CAP does neither of these. Rather, it simply recruits the polymerase to the promoter—it binds cooperatively with polymerase at the gene—and transcription then proceeds. In the absence of any regulator, polymerase occasionally spontaneously binds to the *lac* promoter and initiates transcription. This "basal" level of transcription results from the same enzymatic events and involves the same enzymatic machinery, as does activated transcription; the only difference is that the former occurs less frequently (~40-fold) than does the latter, and so fewer transcripts are made.

Regulated recruitment is thus a quantitative affair: CAP increases the probability of polymerase-promoter interaction and, to that extent, "activates" the gene; Lac repressor decreases the likelihood of that interaction and, to that extent, "represses" the gene.

In general, cooperative binding of proteins to DNA, while stabilizing the protein-DNA interaction, is reversible. If one of the binding proteins is RNA polymerase, however, the reaction becomes essentially irreversible when the polymerase-promoter complex undergoes the transition to the open form.

Each of the protein-protein and protein-DNA interactions encountered in this example involves a patch on the surface of the relevant molecule that we might characterize as "glue-like." This term accurately conveys the image of simple binding (adhesive) interactions. But the interactions are often highly specific and they span a broad range of affinities. Thus, Lac repressor and CAP bind to their respective sites tightly and with high specificities, whereas the interaction of CAP's activating region with polymerase is orders of magnitude weaker.

The signals that govern transcription of the *lac* genes (glucose and lactose) are first detected by the changes (allosteric) they induce in the shapes of the regulatory proteins, promoting DNA binding for CAP and preventing it for Lac repressor. The *meaning* (specificity) of those signals in terms of gene regulation—which genes are transcribed in response to them—is determined by the DNA-binding "address" of those proteins, i.e., where the regulatory proteins bind. Thus, the Lac repressor could be used to bring any gene under the control of lactose merely by inserting its binding site in the promoter of that gene. In nature, it is used at the *lac* genes because these are the genes usefully regulated by lactose.

CAP, in contrast, works on some 200 genes in addition to the *lac* genes. For example, CAP works in conjunction with the Gal repressor to control transcription of the *gal* genes, the products of which metabolize galactose. The ability of CAP to work with disparate regulators (in these cases two different repressors) is an example of combinatorial control.

MORE REGULATED RECRUITMENT: THE BACTERIOPHAGE λ

We noted in the Introduction that the bacterial virus λ can establish dormant residency in an *E. coli* cell. In such a cell, the phage chromosome is

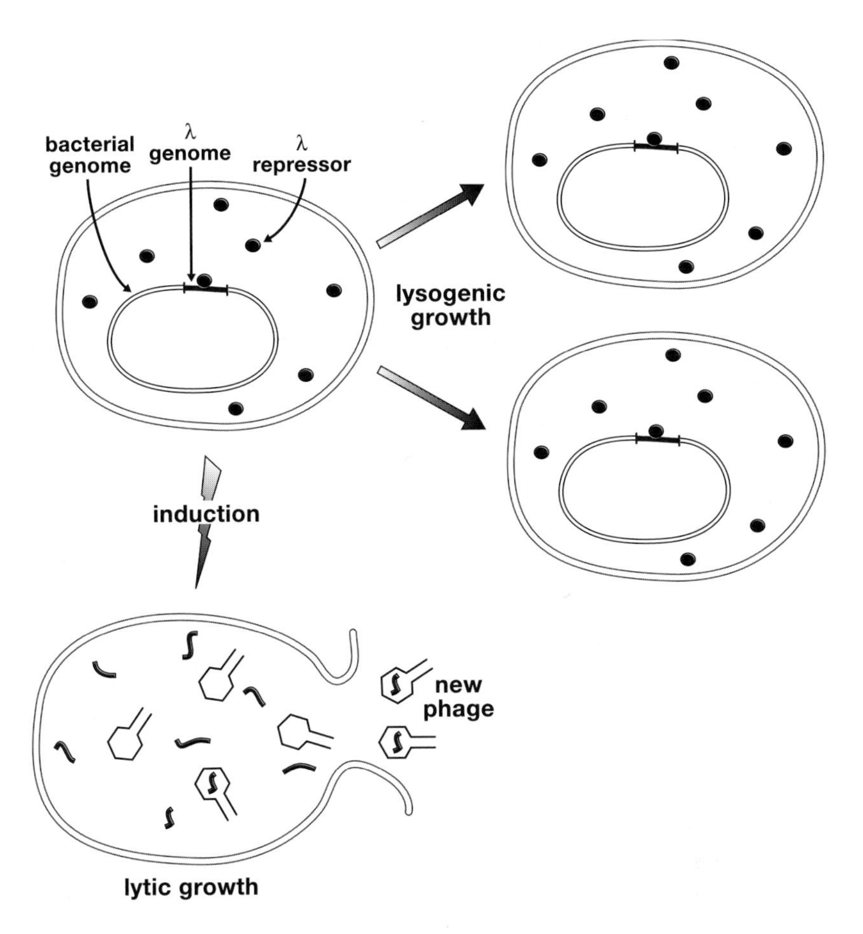

FIGURE 1.10. Growth and induction of a λ lysogen. The phage genome comprises some 50 genes, the majority of which are involved in lytic growth. Following induction, sets of these lytic genes are expressed sequentially.

integrated into—and passively replicated along with—the host chromosome. One phage gene, called *c*I, is expressed: the product of that gene, the bacteriophage λ repressor, keeps other phage genes switched off. Upon exposure to UV light, the repressor function is abolished, with two consequences: the previously inert phage genes (called lytic genes) are switched on, and the repressor gene itself is switched off. Some 45 minutes later, the cell lyses to release a crop of about 100 new phage particles (Figure 1.10).

A bacterium carrying a dormant phage chromosome is called a lysogen; the switch to lytic growth is called induction.

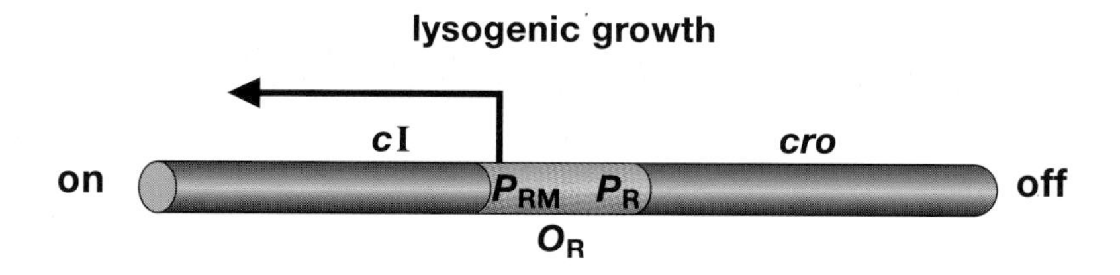

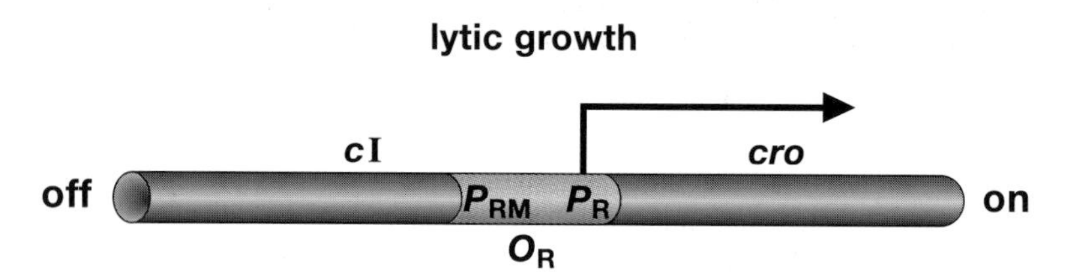

FIGURE 1.11. The biphasic switch. O_R has its name because there is another control region, located on the left side of *c*I, called O_L. That operator controls a second set of lytic genes. The term O_R is shorthand for a region comprising three operator sites and two promoters as shown in Figure 1.12. Because the repressor (*c*I) and *cro* genes are transcribed in opposite directions, they are transcribed off different strands of the DNA.

The Switch

To understand how the switch works, we need consider two regulatory genes (*c*I and *cro*) and the regulatory region called O_R (right operator), which are shown in Figure 1.11. In a lysogen, *c*I is on and *cro* is off, and vice versa when lytic growth ensues. Figure 1.12 shows this regulatory region in more detail. The operator comprises three binding sites—O_{R1}, O_{R2}, and O_{R3}—that overlap two opposing promoters. One of these, P_R, directs transcription of lytic genes and the other, P_{RM}, directs transcription of the *c*I gene.

Figure 1.13 shows that in a lysogen, the λ repressor (the product of the *c*I gene as we have said), at O_R, is bound mainly at the two adjacent sites O_{R1} and O_{R2}. At these positions, it performs two functions: it represses rightward transcription from the promoter P_R, thereby turning off expression of *cro* and other lytic genes; simultaneously (and despite its name) it activates transcription of its own gene from the promoter P_{RM}.

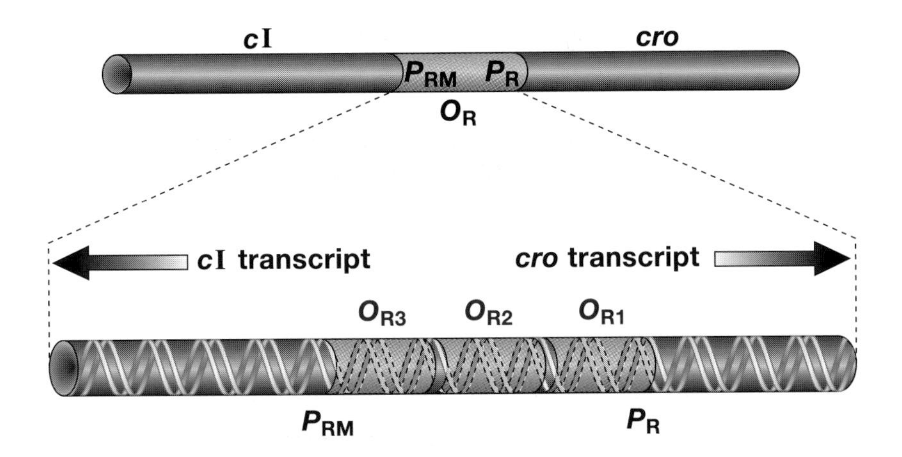

FIGURE 1.12. Arrangement of genes and sites at the O_R region of λ. The three operator sites are each 17 bp. O_{R1} lies within P_R, and O_{R3} lies within P_{RM}. P_R is a strong promoter that functions efficiently without an activator, whereas P_{RM} requires an activator to work efficiently. Both transcripts are shown, but please recall that, as indicated in Figure 1.11 and in the text, only one or the other gene is expressed at any given time. (Modified, with permission, from Ptashne 1992.)

Upon induction, repressor vacates the operator and transcription from P_R—an inherently much stronger promoter than P_{RM}—commences spontaneously. The first newly made protein is Cro: This protein binds first to O_{R3}, apparently helping to abolish repressor synthesis.[10]

Establishing Lysogeny

According to the description given thus far, the λ repressor is required to activate transcription of its own gene. This raises the question of how expression of that gene is turned on when the virus first infects a bacterium to establish lysogeny. Figure 1.14 shows the answer: the repressor gene is initially transcribed from a promoter called P_{RE} (*p*romoter for *r*epressor *e*stablishment).

Transcription from P_{RE} is activated by the product of another phage gene, *c*II. The newly made repressor (CI) activates transcription of its own gene from P_{RM} (*p*romoter for *r*epressor *m*aintenance) as it turns off transcription of other phage genes, including that of *c*II. The lysogenic system thus established is then self-perpetuating in the absence of an inducing signal.

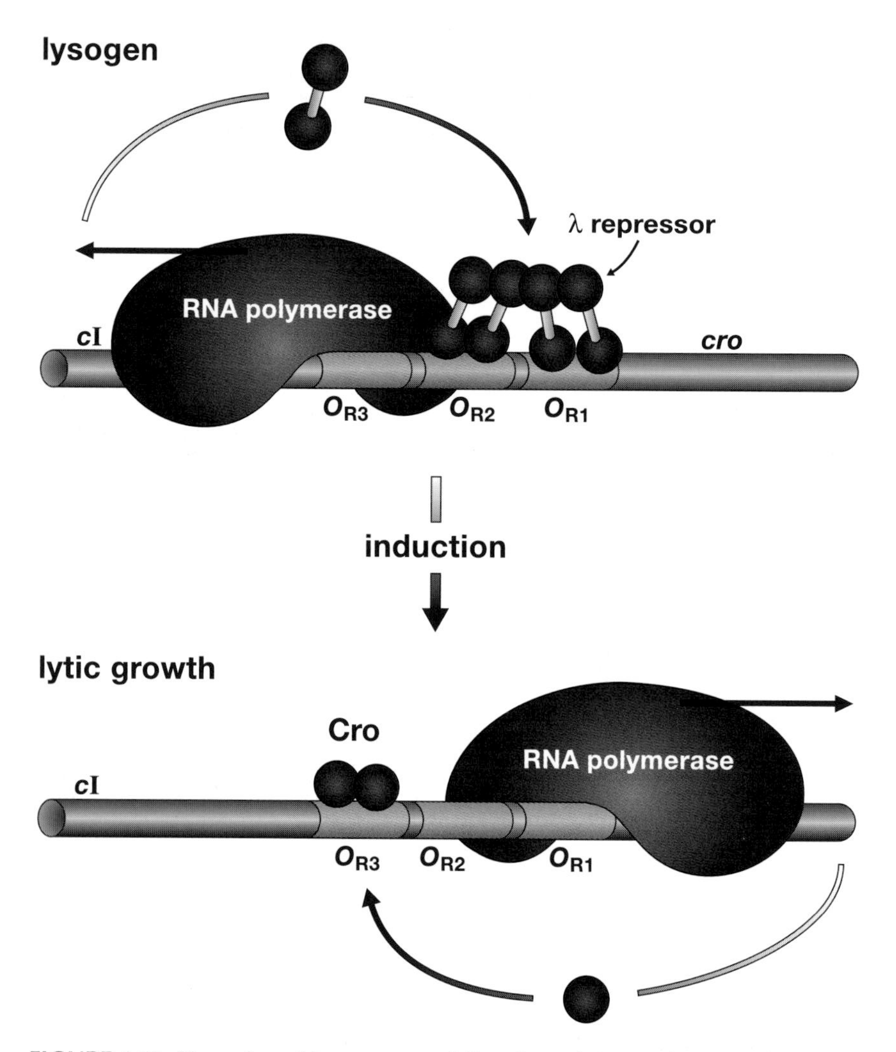

FIGURE 1.13. The action of λ repressor and Cro. O_{R1} and O_{R2} overlap P_R, and repressor bound to those sites turns off *cro* and the other lytic genes. O_{R2} is positioned such that repressor bound there contacts polymerase at P_{RM} and thereby activates the *c*I (repressor) gene. O_{R3} lies within P_{RM}, and Cro bound there abolishes transcription of *c*I. (Modified, with permission, from Ptashne 1992.)

Bacteriophage λ does not always lysogenize when it infects—sometimes it grows lytically in newly infected bacteria. The choice between establishing lysogeny and growing lytically is, like the process of induction, influenced by extracellular signals (see below, Detecting Physiological Signals).

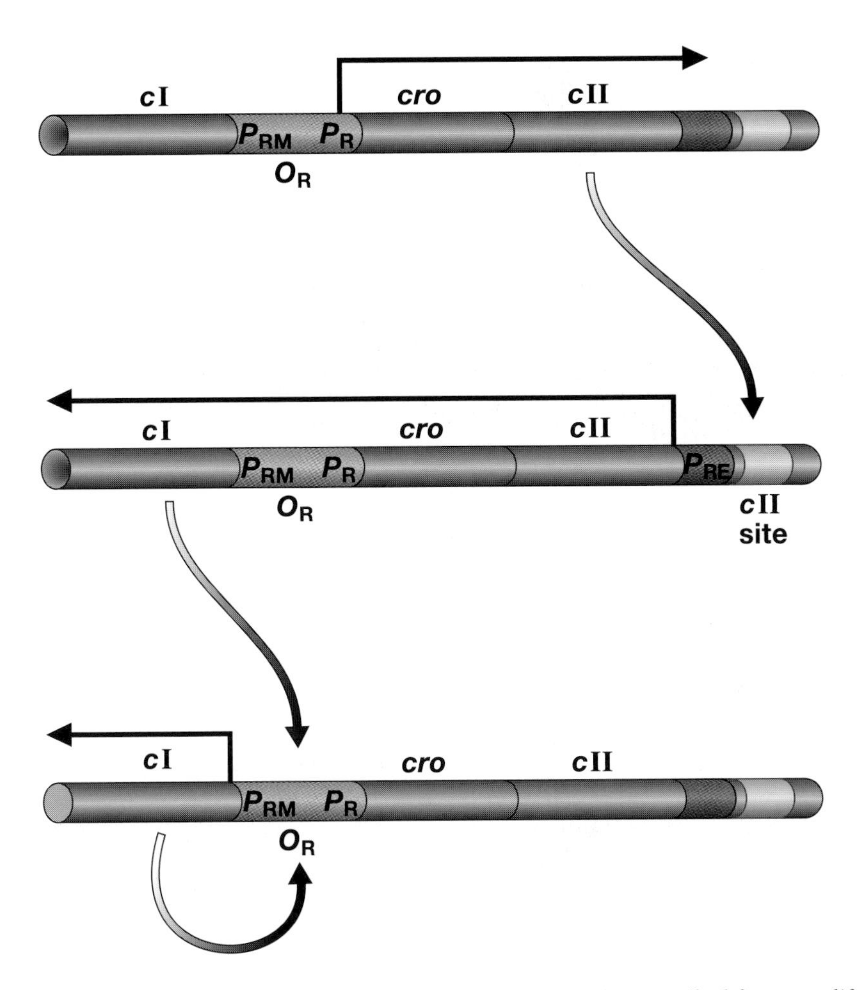

FIGURE 1.14. Establishment of lysogeny. The same gene, *c*I, is transcribed from two different promoters: from P_{RE} to establish lysogeny and from P_{RM} to maintain that state. Repressor bound at O_{R1} and O_{R2} turns off the establishment mode of expression (which depends on transcription from P_R) as it activates the maintenance mode (transcription from P_{RM}). As implied by this figure, P_R controls not only lytic genes (as indicated in the text), but also *c*II, which is required to establish lysogeny. Similarly, P_L, which controls many lytic genes, also controls a few genes which help establish lysogeny.

Analogies with *lac*

The essential features of gene regulation as described for the *lac* case are also represented in the λ switch.

Promoters

Of the three promoters in the λ switch (P_R, P_{RM}, and P_{RE}) two, P_{RM} and P_{RE}, resemble the wild-type *lac* promoter: they are weak and require activators to work efficiently. Promoter P_R, in contrast, functions at a high level in the absence of any activator.

Protein-DNA Interactions

λ repressor, Cro, and CII all bind to DNA using the familiar HTH motif. In each case, residues along the recognition helix, as well as certain other residues, direct the protein to its specific site or sites on DNA.

Repression

Repression is effected by a mechanism essentially the same as that encountered in the *lac* case. Thus, in a lysogen, λ repressor excludes polymerase from P_R, and upon induction—as lytic growth begins—Cro excludes polymerase from P_{RM}.

Activation

Again, as in the *lac* case, a protein-protein interaction mediates gene activation. Thus, repressor bound at O_{R2} contacts polymerase at P_{RM} to stimulate repressor synthesis in a lysogen. The same rule applies for CII: this protein, bound to a site adjacent to P_{RE}, contacts polymerase at that promoter to stimulate repressor synthesis.

Experiments similar to those that we described for *lac* show that polymerase need only be recruited to P_{RM} to achieve high levels of transcription. For example, activator bypass experiments work at P_{RM} as they do at the *lac* promoter. The following are a few additional experiments characterizing activation in this case.

- ***Polymerase target mutants.*** *pc* mutants of λ repressor, like *pc* mutants of CAP, bind DNA but do not activate transcription. Starting with such a *pc* mutant, a mutant of RNA polymerase was found that restored activation. The mutation results in an amino acid change in σ and is believed to identify the surface contacted by repressor's activating region.

- ***Creating a new activator.*** Cro works only as a repressor: even when positioned adjacent to a promoter, it fails to activate transcription because it lacks an activating region. Residues on one surface of Cro were changed so as to resemble the activating region of the λ repressor. When bound near P_{RM}, the modified Cro activated transcription from that promoter. Thus, an activating region can work when presented in different protein contexts, provided the DNA-binding address is appropriate.
- ***Activating region variants.*** Many variants of the λ repressor's activating region have been generated artificially. Among those that retain function, some work weakly and others work up to about fivefold better than the wild type. The ease with which these variants are found reinforces the notion that an activating region need only provide a simple binding surface, rather than performing some more specialized function.
- ***CAP activation of P_{RM}.*** On a DNA template bearing P_{RM} and a suitably positioned CAP-binding site (see Figure 1.2), CAP activates P_{RM}. Thus, the interaction that works at the *lac* promoter also works at P_{RM}.

Detecting Physiological Signals

In the *lac* case, the relevant physiological signals are communicated by the binding of two small molecules (cAMP and allolactose) to the regulatory proteins CAP and Lac repressor, respectively. In the λ case, there are two stages at which physiological signals impinge. In both cases, the signals trigger proteolysis of regulatory proteins, λ repressor in one case and CII in the other.

Upon infection, the decision of whether or not to establish lysogeny is determined by the activity of a host-encoded protease that attacks CII. When the cells are growing vigorously, the protease is particularly active. CII is destroyed and repressor synthesis does not initiate, and so the infecting phage usually chooses lytic growth. If the cells are growing poorly, the protease is relatively inactive and so, because CII accumulates to a high level, the infecting phages readily establish repressor synthesis and usually lysogenize the host.

Inducing agents, such as UV light, damage DNA. The damaged DNA

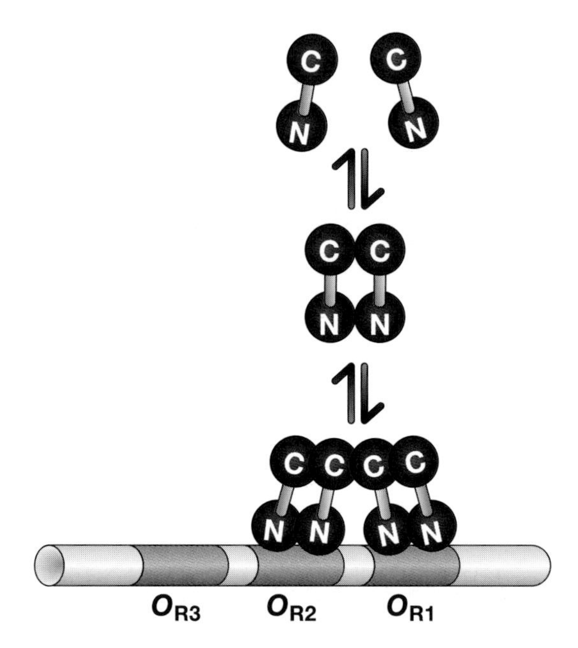

FIGURE 1.15. Cooperative binding of λ repressor to DNA. The λ repressor amino domain (N) is separated from the carboxyl domain (C) by a linker of 40 amino acids. Repressor binding to these sites would be further helped by the interaction between repressor at O_{R2} and polymerase at P_{RM}.

binds to and activates the protein RecA. Activated RecA interacts with and stimulates cleavage of repressor, resulting in lytic growth.[11]

Making an Efficient Switch

The λ switch is highly efficient. In the absence of an inducing signal, spontaneous induction is rare (less than once in a million cell divisions). But, when exposed to an appropriate signal, virtually every cell in the lysogenic population produces phage. Various features contribute to this efficiency; none requires interactions fundamentally different from those we have already encountered, as we now show.

Cooperative Binding of λ Repressor to DNA

λ repressor binds cooperatively to the operator sites O_{R1} and O_{R2} as shown in Figure 1.15. Monomers first interact to form dimers, and then two dimers interact to bind to the operator sites as shown.[12]

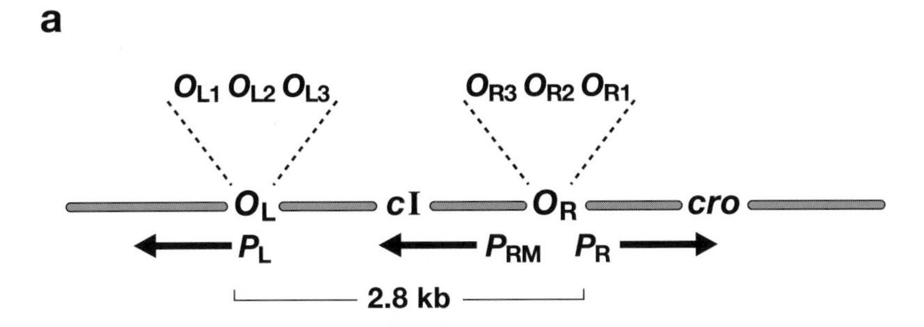

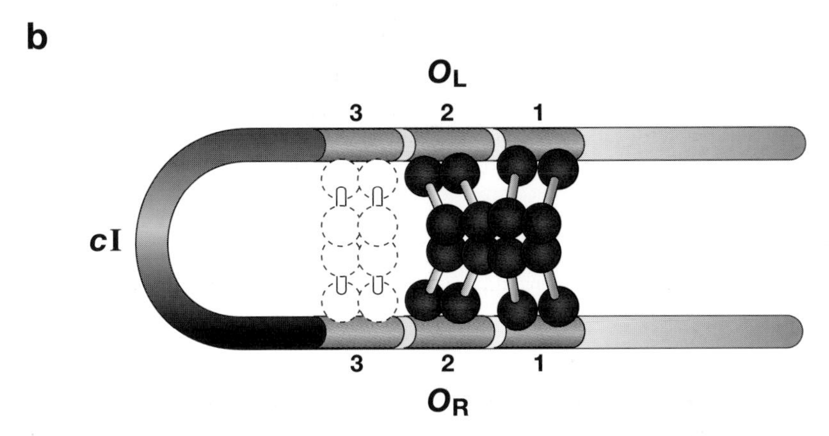

FIGURE 1.16. Interaction of repressors at O_R and O_L. Interaction between repressor at O_R and O_L stabilizes binding and increases repression at O_R (and presumably at O_L). The crystallographic structure of the carboxyl domains of repressor suggests that it could form an octamer, and such octamers have been observed at high repressor concentrations in vitro.

The repressor tetramer bound at O_R as in Figure 1.15 undergoes a further interaction: an octamer is formed by the interaction with another pair of dimers bound to the second major λ operator O_L (left operator). This operator is positioned about 3 kb away from O_R and controls the promoter P_L which, in turn, directs synthesis of another group of λ genes. The interaction of repressors at O_R and O_L is accompanied by formation of a large DNA loop as observed in vitro and shown in Figure 1.16.

This long-range interaction between repressors at O_L and O_R has two effects: it increases repression at P_L and P_R a fewfold and, as we describe below, increases the negative autogenous (self) regulation of repressor.

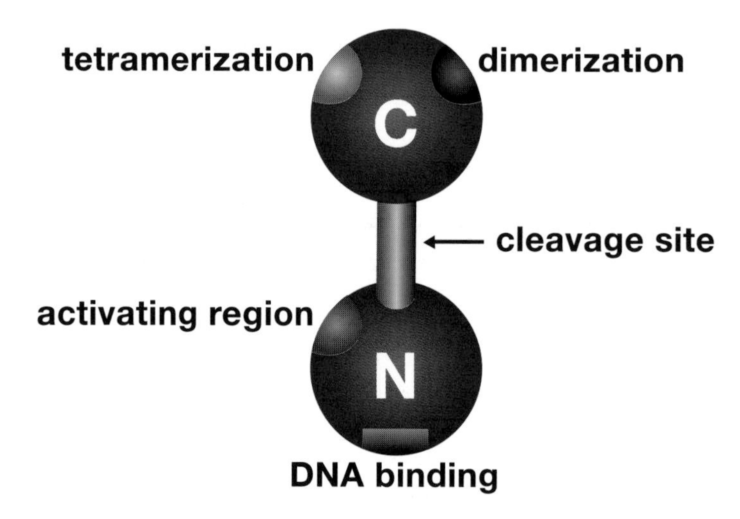

FIGURE 1.17. The functional patches on λ repressor. The activating region contacts polymerase, and "tetramerization" indicates the patch at which two dimers interact when binding cooperatively to sites on DNA. These same patches also mediate octamerization. The "cleavage site" shows where repressor is cleaved upon induction. As before, N indicates the amino domain.

A consequence of these interactions is that the curve describing the binding of repressor to the two operator sites, as a function of the repressor concentration, is highly inflected, or "sigmoid." This means that, around a certain value, small changes in repressor concentration can have large effects on site occupancy.[13]

Cooperativity has a crucial role in the λ switch, as revealed by the mechanism of induction: the inducing agent simply eliminates cooperative binding of repressor monomers to DNA. To see this, we must describe the positions of the various functional patches on the repressor's surface, and the fate of repressor upon induction.

As shown in Figure 1.17, both the DNA-binding surface and the activating region are carried on the amino domain of the repressor. In contrast, the cooperativity functions of repressor—those surfaces that mediate formation of dimers and interaction between dimers—lie on the carboxyl domain. Cleavage of λ repressor upon induction, mediated by RecA, results in separation of the amino from the carboxyl domain.

The separated amino domain is capable of binding DNA and activating transcription, but as it cannot bind *cooperatively*, its concentration in the cell is too low to maintain lysogeny. When artificially expressed at high concentrations, the separated amino domains bind DNA and activate transcription.

Elimination of cooperativity between λ repressor monomers has the same effect as decreasing the concentration of repressor monomers some 100-fold. Because only a 5–10-fold decrease in concentration is required for repressor to completely vacate the operator, induction proceeds efficiently.

Autogenous Control by Repressor

To ensure that binding of repressor depends on cooperativity—and the distinct "all-or-none" nature of the switch is maintained—the concentration of repressor in the lysogen must be held at a suitably low level. The design of the switch ensures that repressor concentration never exceeds that level by the following form of autogenous negative control. In addition to sites O_{R1} and O_{R2}, repressor can also bind to the third site, O_{R3}. But O_{R3} is an intrinsically weak binding site for repressor and, moreover, repressors bound at O_{R1} and O_{R2} do not bind cooperatively with repressor at O_{R3}.[14]

Site O_{R3} is thus filled only if the repressor dimer concentration increases to a level higher than that required to fill O_{R1} and O_{R2}. When bound to site O_{R3}, the repressor shuts off synthesis of further repressor by excluding polymerase from P_{RM}. This prevents the synthesis of new repressor until the existing repressor concentration falls to the appropriate level (a consequence of dilution through cell division).

In the absence of this form of autogenous negative control (e.g., if O_{R3} is mutated so that repressor cannot bind there), the level of repressor in the lysogen increases some threefold, and induction is severely impaired.

Repressor binding to O_{R3} is improved by interaction of repressors bound at O_R with those bound at O_L. A likely explanation for this effect is found in Figure 1.16. There we see a loop formed by an octamer of repressor bound simultaneously at O_{R1}, O_{R2}, O_{L1}, and O_{L2}. Under these circumstances, O_{L3} and O_{R3} are held in close proximity. This configuration evidently helps another pair of repressor dimers bind cooperatively to O_{L3} and O_{R3} (as indicated by the dotted repressors in the figure).[15]

Induction is further facilitated by the loss of autogenous positive control by repressor. As repressor occupancy of the operator site O_{R2} falls, the rate of synthesis of new repressor also falls, because repressor no longer stimulates transcription of its own gene. The drop in repressor concentration is thus more dramatic than it would otherwise be.

An important form of gene regulation that does not involve control of transcriptional initiation is also found in λ and is briefly described in the Antitermination panel.

ANTITERMINATION

In the text we stress the action of regulators that activate or repress the initiation of transcription. Here we note that a subsequent step—elongation of the transcript—can be regulated as well. Thus the polymerase, having initiated transcription, pauses when it encounters certain DNA sequences, and terminates (and falls off the DNA) at others. We know of two kinds of "antiterminators" encoded by bacteriophage λ that cause the polymerase to read through such signals.

λ *Q*

The product of the phage *Q* gene is a DNA-binding protein that recognizes a DNA sequence (called *qut*: *Q utilization*) positioned between the –10 and –35 regions of a promoter called P'_R. This promoter is responsible for directing transcription of genes expressed late in the life cycle of the virus. In the absence of *Q*, polymerase initiates transcription at the strong promoter P'_R, pauses 17 bp downstream, and then terminates (and falls off the DNA) at a terminator positioned some 200 bp downstream. *Q*, made midway in the lytic cycle, binds to *qut* and interacts with the paused polymerase. Polymerase now reads through the pause site and—evidently because *Q* has joined the polymerase—reads through the downstream terminator. In fact, the transcript extends more than 26,000 bp, ignoring numerous other termination sites on route.

λ *N*

The product of the phage *N* gene is an RNA-binding protein that recognizes the *nut* (*N utilization*) site present in two phage mRNAs. These mRNAs emanate from P_R and P_L; in the absence of N protein, the former terminates after transcribing *cro*, and the latter terminates after transcribing *N*. N binds *nut* and polymerase and evidently joins the enzyme to make it impervious to the terminators. N is helped in this regard by four other bacterial proteins, including one called NusA. *nut* (as a DNA sequence) was inserted between other promoters and terminators and, when provided, N caused antitermination in those cases as well.

Interim Summary and Extensions

In the λ switch, we encounter several features not found in the *lac* case. These include a regulator (λ repressor) that works as an activator and as a repressor; cooperative binding of that regulator to adjacent operator sites; binding of different regulators (repressor and Cro) to the same operator sites but with different orders of affinities and with different effects; autogenous positive (λ repressor) and negative (λ repressor and Cro) control; and the use of two promoters (P_{RE} and P_{RM}), one transiently, to direct transcription of the same gene (the repressor gene). However, none of these added features requires molecular interactions other than those of the kind we encountered at *lac*.

For example, there is nothing particularly surprising about the ability of the λ repressor to work both as an activator and as a repressor. The repressor-binding sites are arranged so that, in a lysogen, the repressor is positioned to exclude polymerase from the promoter of lytic genes and to touch, with its activating region, polymerase at P_{RM}. CAP, which at *lac* works only as an activator, also works as a repressor at certain other promoters, depending on the location of its binding site on DNA.

The cooperativity between λ repressors, which helps render the switch so efficient, has been imposed in a simple way. Instead of having one site to which the activator—λ repressor—binds, there are two sites to which it binds cooperatively. That cooperative binding in turn depends on adhesive interactions between the repressor molecules. Induction separates the cooperativity domain from the DNA-binding domain, and that loss of cooperativity is sufficient to flip the switch.

It is apparent that the same kinds of interactions—binding interactions that we have characterized as adhesive—assorted and reiterated in different ways, can be used to solve quite distinct gene regulatory problems.

Activation: A Closer Look

The two activators that we have described—CAP working at the *lac* genes, and λ repressor working at the λ *c*I gene—recruit polymerase by contacting different parts of it. Thus, CAP contacts the carboxyl domain of α, and λ repressor contacts σ. There is nothing inherent in the promoters that requires these specific interactions: CAP will activate the λ *c*I gene if its binding site is suitably positioned.

Other activators that work by recruitment (and even CAP working at different promoters) contact various different sites on the carboxyl domain of α, the amino domain of α, and σ. Thus, although a stereospecific complex of activator, polymerase, and DNA is (presumably) formed in each case, there are many arrangements that will suffice for activation. In addition, as revealed by the activator bypass experiments, even "artificial" interactions between a DNA-binding protein and polymerase will work.

We have noted that polymerase binding to any of the promoters we have discussed thus far involves two steps: polymerase forms a transient closed complex with the DNA, which then, with some probability, undergoes a transition to a stable open complex in which the DNA strands around the promoter have separated. Because the closed complex is unstable, recruitment is not accomplished until the open complex is formed. Activators could, in principle, affect either step to achieve recruitment. And indeed, measurements performed in vitro suggest that whereas a given activator working at one promoter affects primarily the first step, another activator working at another promoter affects the second step.

For example, λ repressor is reported to affect the second (isomerization) step at P_{RM}, and CAP is reported to affect the initial binding of polymerase at the *lac* genes. We do not believe, however, that these differences require different kinds of activator-polymerase interactions. Rather, both can be accounted for by simple adhesive interactions between activator and polymerase as we now explain.

The detailed geometries of the tripartite complexes of activator, polymerase, and DNA determine which step (of the two described above) is stimulated. Thus, were the contact between activator and polymerase to be favored when the polymerase and DNA formed a closed complex, the initial binding would be stimulated. Were the activator-polymerase contact favored when the polymerase and DNA formed some intermediate between the closed and open complexes, then a step subsequent to initial binding would be stimulated.[16]

Perhaps what evolution selected in each of these cases was the positioning of an activator so that it could touch polymerase, using a promoter and polymerase design that has allowed many different interactions to be productive. Inevitably, some such interactions will affect one step of the reaction, and some interactions will affect the other. We would expect, of course, that not every contact between a DNA-tethered protein and polymerase would lead to gene activation. Presumably, only certain geometric arrangements of the polymerase-promoter-activator complex will be productive, and some could indeed even be inhibitory (see panel on More on Repression in Bacteria at the end of this chapter).

DNA Binding: A Closer Look

For activators that work by recruitment (or for repressors that counter their effects), the detailed mode of DNA binding is not relevant to their functions. What is required is that the proteins bind with the appropriate affinities to their specific sites (locations) on DNA. Even for proteins that use the same motif (e.g., the HTH motif), the details of DNA binding vary.

For example, whereas the λ repressor and Cro have virtually no effect on the helical structure of DNA, a closely related repressor from phage 434 contorts the center of its operator, and the ability of the operator to be so deformed, a feature dependent on its sequence, affects the affinity of the operator for repressor. To take another example, CAP dramatically bends the DNA. This bending is the result of the DNA partially wrapping around the protein, a contortion that is imposed by protein-DNA interactions beyond those made by the HTH motif. Those additional contacts increase the affinity of CAP for DNA, but they are not required for activation. Thus, CAP mutants that cannot bend DNA can, nevertheless, activate at *lac* normally, but only when present at sufficiently high concentration to bind DNA in the absence of the additional contacts.

We will see below a counter-example in which a change in DNA structure—and therefore the mode of DNA recognition—is key to activation (see below, MerT).

Synergy

We say that two activators work synergistically when the pair elicit more activity than the sum of the activities elicited by each working alone. λ repressors work synergistically, as we have seen, by helping each other bind to DNA. But activators that work by recruitment can work synergistically even if they do not directly interact with each other. Thus, two or more DNA-bound activators, by simultaneously contacting different sites on the same polymerase molecule, can work synergistically: each can contribute to the binding energy required to recruit that enzyme.

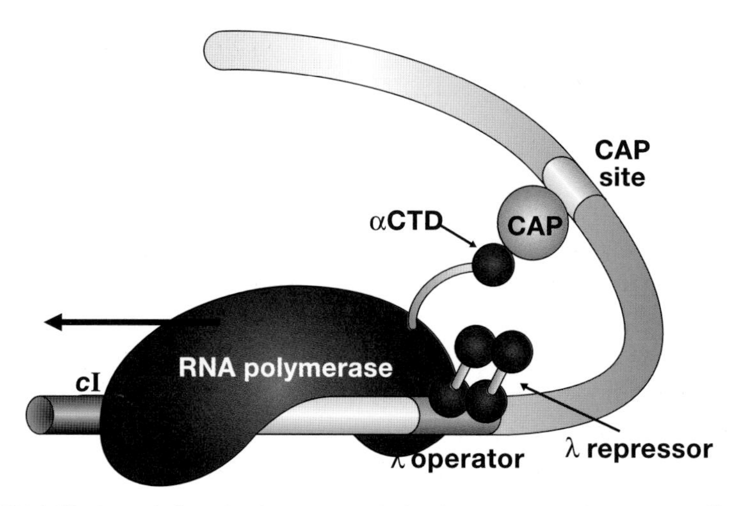

FIGURE 1.18. Synergistic activation. A CAP site has been positioned upstream of λ's P_{RM}. λ repressor bound at its ordinary site O_{R2} contacts σ; CAP bound upstream contacts the αCTD. As in Figure 1.7, only one CAP monomer and one αCTD are shown. Together, the activators increase transcription about 100-fold, whereas each separately has only about a 10-fold effect on this template.

An example of this effect is shown in Figure 1.18: CAP and λ repressor work synergistically when both are bound, in suitable positions, upstream of a gene. This form of synergy readily lends itself to combinatorial control: in principle, as long as activators can contact different sites on the polymerase, they can work synergistically in different combinations.[17]

In the following sections, we describe two examples of gene activation that illustrate mechanisms different from regulated recruitment. One of these involves a polymerase bearing a σ subunit (σ^{54}) other than the most common, σ^{70}; the other involves the common form of polymerase but working at unusual promoters.

POLYMERASE ACTIVATION: *glnA* AND RELATED GENES

The *glnA* gene encodes an enzyme (glutamine synthetase) involved in biosynthesis of the amino acid glutamine. Transcription of this gene is activated by NtrC, a DNA-binding protein that works in response to nitrogen starvation. The *glnA* gene is transcribed by a polymerase bearing the σ^{54} subunit. Approximately 1–2% of the *E. coli* genes are regulated by NtrC.

Promoter Recognition by σ^{54}-containing Polymerase

Not only does the σ^{54} polymerase recognize promoters different from those recognized by σ^{70}-containing polymerase, it does so in an importantly different fashion. Thus, in vivo, in the absence of activator, the polymerase is found bound to the promoter in a stable *closed* complex (see Figure 1.19a). Similarly, in vitro, the enzyme binds the promoter but remains in the closed complex in the absence of activator. Spontaneous isomerization to the open complex occurs so rarely as to render basal transcription virtually nonexistent. Even at high concentration of enzyme—readily achieved in vitro—no transcription is observed in the absence of the activator. Thus, unlike in the *lac* and λ cases, recruitment does not suffice for activation: the enzyme is stably bound (i.e., is already recruited) in the absence of the activator.[18]

DNA Binding by NtrC

Four adjacent NtrC-binding sites are positioned about 150 bp upstream of the transcription start site. Each site is recognized by an NtrC dimer, and each dimer recognizes its site using a variant of the HTH motif we have encountered previously. The dimers bind highly cooperatively.

Detecting the Physiological Signal

The absence of nitrogen is sensed by the NtrB protein, which responds by phosphorylating NtrC. Phosphorylated NtrC, in turn, works as an activator. There are mutants of NtrC that work constitutively in the absence of phosphorylation.

Activation by NtrC

NtrC activates transcription by contacting the polymerase prebound to the promoter. Unlike the regulated recruitment mechanism, activation in this case requires hydrolysis of ATP. The activator contacts the σ^{54} subunit and, as it hydrolyzes ATP, induces a conformational change in polymerase. That change triggers isomerization of the polymerase-DNA complex to the open form, and thus the initiation of transcription. The activating-region–polymerase interaction in this case is “instructive,” not merely adhesive.

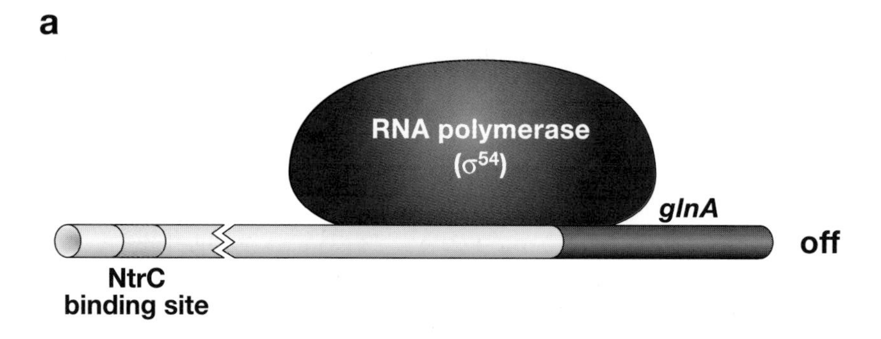

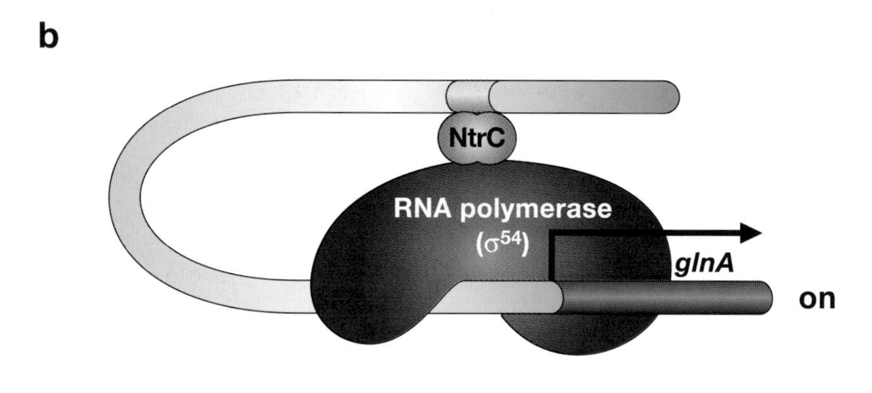

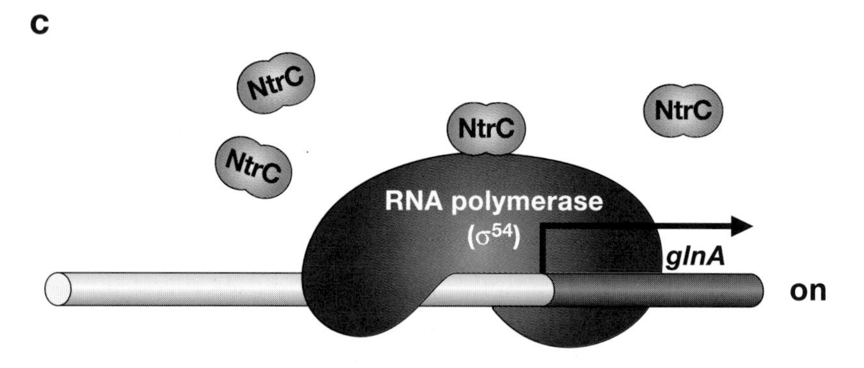

FIGURE 1.19. σ^{54}-containing polymerase and activation by NtrC. The sequence elements recognized by σ^{54}-containing holoenzyme are different from those recognized by σ^{70}-containing holoenzyme. Instead of those shown in Figure 1.5, we have variants of the sequence ttGGcacaNNNNttGCA (where capitalized, bases are strongly preferred; where in lowercase, bases are somewhat preferred; N indicates positions where there is no preference). Although not specified in the figure, the "privileged site" that NtrC must contact on this polymerase is within the σ^{54} subunit.

Because the activator binding sites are not immediately adjacent to the promoter, the intervening DNA loops out to accommodate the activator-polymerase interaction (see Figure 1.19b). This looping out is particularly evident if the activator sites are moved even further away, as far as 1–2 kb upstream. In that case, the looping is readily visualized by electron microscopy. At some NtrC-activated genes, a protein (IHF) binds between the activator and the gene, bending the DNA. This bending, if in the proper phase, promotes activation by facilitating looping.

In contrast to the action of CAP and λ repressor, gene activation by NtrC is not absolutely dependent on the activator's ability to bind DNA. Thus, if the DNA-binding function is destroyed by mutation, or if there are no NtrC-binding sites near the gene, NtrC will nevertheless activate, but only if present at an abnormally high concentration (see Figure 1.19c). This is because the role of the activator is not to recruit the polymerase, but rather to act on an already bound polymerase. The ordinary role of DNA binding of NtrC is to increase the concentration of its activating region in the vicinity of the promoter-bound RNA polymerase. When overexpressed, NtrC activates all genes bearing a prebound σ^{54} polymerase, regardless of the presence of NtrC-binding sites.

Other Activators of σ^{54} Polymerase

NtrC is not the only activator that works on genes transcribed by RNA polymerase bearing σ^{54}. For example, the activator XylR responds to xylene and controls the genes required for xylene degradation, and the activator DctD, in response to dicarboxylic acids, activates genes encoding proteins that transport those acids into the cell. These NtrC-like activators were identified in a variety of bacterial species, and they all work when transferred to *E. coli.*

Each of these activators works specifically at the appropriate genes as determined by its DNA-binding address. Thus, for example, if the NtrC-binding sites are replaced by binding sites for XylR, that activator now works on the *glnA* gene, and does so in response to xylene.[19]

Interim Summary

Activators of the class represented by NtrC work quite differently from those represented by CAP and λ repressor. Whereas CAP and λ repressor need only recruit a constitutively active form of polymerase (bearing σ^{70}) to the gene, NtrC must literally activate a polymerase (bearing σ^{54}) pre-bound to the promoter. These different requirements have the following consequences:

- Activators that work simply by recruitment, e.g., CAP, must bind DNA to work. In contrast, activators of the NtrC class can work without binding DNA, but only if their concentration is increased to artificially high levels. At such high concentrations, all genes bearing σ^{54}-bound polymerases are activated, and thus the specific link between signal and gene is lost. At the ordinary concentration of NtrC in the cell, DNA binding imposes specificity: it increases the concentration of the activating region near the gene to be activated.

- The activating domains of activators that work by recruitment need only present a glue-like surface that interacts with polymerase, and there are many sites on polymerase that can be fruitfully touched. In contrast, activators such as NtrC interact with the enzyme in a more specialized way to induce the required conformational change; in that case, there is, as far as we know, a single privileged site on polymerase that must be contacted.

- One form of synergistic activation we described for activators that work by recruitment does not occur for activators of the NtrC class. That form of synergy required that two DNA-tethered activators simultaneously interact with different surfaces of polymerase. But for the NtrC case, because each activator touches the same site, two activators of this class cannot touch the enzyme simultaneously.

- The typical gene transcribed by polymerase bearing σ^{54} has essentially no basal level of expression and a high activated level. This range of expression is made possible by using the energy of ATP to convert an inert enzyme into an active one. In contrast, genes that are read by the σ^{70} polymerase are controlled, as we have seen, solely by limiting or increasing access of the polymerase to the gene. In that case, the same wide range of expression is typically achieved by using both an activator (to turn it on to a high level) and, in the absence of the activator, a repressor (to turn it off to a low level).

Analysis of NtrC and related activators not only reveals a mechanism of activation different from recruitment, but also illustrates how proteins binding to widely separated sites on DNA can interact. That is, if the interaction is sufficiently strong, the DNA loops out to accommodate that interaction. The looping has no direct role in activation: as we have seen, activation can even be achieved in the absence of DNA binding. We know of one case in which an activator moves along the DNA to get to the promoter, an effect that requires concomitant DNA replication. This is illustrated in the panel Phage T4 Late Genes.

PHAGE T4 LATE GENES

This example shows a way, other than by DNA looping, that an activator, initially bound at a distance from a promoter, can communicate with that promoter.

T4 is a phage that infects *E. coli* but, unlike λ, it only grows lytically. Lytic growth of T4 requires the sequential expression of sets of genes, the last of which comprise the late genes. The promoters of the late genes are recognized by *E. coli* polymerase bearing yet another σ factor, σ^{55}, the product of a T4 gene expressed earlier in the life cycle. Expression of these late genes is linked to replication of the phage genome.

This mechanism involves a moving activator. The machinery that replicates T4 DNA—like all DNA replication machines—includes a "sliding clamp" protein. This protein encircles and moves along the DNA, ensuring processivity of replication. In the T4 case, this sliding clamp interacts with RNA polymerase at the late gene promoters to activate transcription. It is not known whether that interaction simply recruits polymerase or performs some more elaborate function to stimulate transcription.

PROMOTER ACTIVATION: *merT* AND RELATED GENES

The *merT* gene encodes a protein that protects bacteria against mercury poisoning, and transcription of the gene is induced by the presence of mercury. The gene is read by polymerase bearing σ^{70} and is controlled by the activator MerR.

Promoter Recognition

The *merT* promoter is unusual in that the –10 and –35 regions are separated by 19 bp instead of the usual 17 bp, and thus the two crucial regions of the promoter are not properly spaced or aligned. In the absence of mercury, polymerase binds to the promoter, along with the regulator MerR as shown in Figure 1.20, but cannot initiate transcription. MerR binds near the middle of the promoter on the face opposite that bound by polymerase.

MerR works as a repressor in the absence of mercury. It holds the promoter in the configuration unsuitable for transcription. Thus, in a mutant bacterium lacking MerR, but not in wild-type cells, there is a detectable basal level of transcription.

Detecting the Physiological Signal and Activation

MerR undergoes a conformational change upon binding mercury that in turn twists the promoter, bringing together and properly aligning the –10 and –35 regions. This new alignment allows efficient transcription. Constitutive mutants of MerR twist the DNA even in the absence of mercury, and deletion mutants of the *merT* promoter, which lack two base pairs, are transcribed at a high level in the absence of MerR.

Several other *E. coli* genes are controlled by regulators similar to MerR. These genes, like *merT*, are activated by toxic metals, and their products protect against those metals. ZntR controls a gene whose product protects against zinc; CueR regulates a gene whose product protects against copper. In each case, the activator recognizes the DNA sequence at the relevant promoter and, on binding the particular metal, twists that promoter into an active state.

Interim Summary

Unlike the previous two types of activation represented by CAP (recruitment) and NtrC (polymerase activation), the precise mode of DNA binding *is* crucial for the action of MerR. There is no separate activating region, and apparently no contact with polymerase is necessary for activation; rather, the effect of DNA binding on promoter structure is the basis of activation. This contrasts with the previous cases in which the function of DNA binding was simply to "locate" the activator near the gene to be activated.

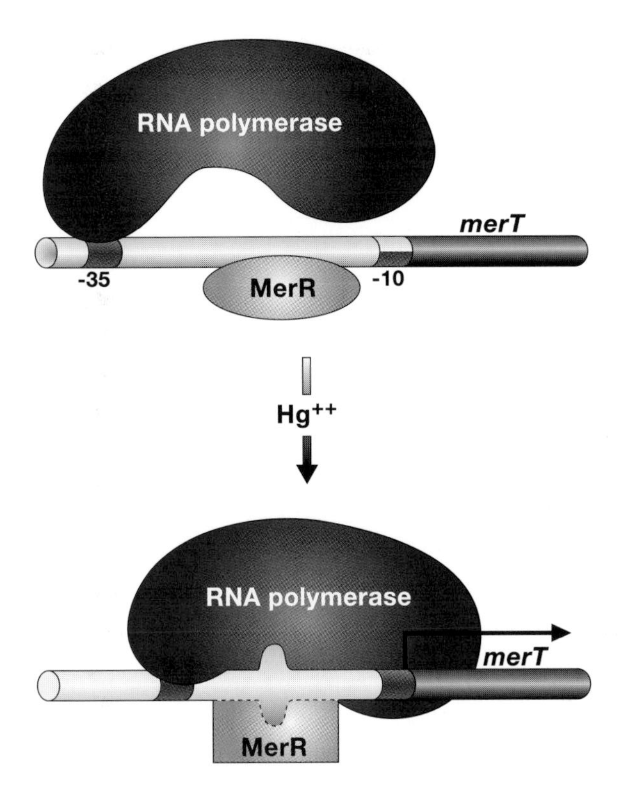

FIGURE 1.20. Activation by MerR. The promoter bears the –10 and –35 elements on nearly opposite sides of the helix. In the absence of mercury, MerR binds with polymerase to form an inactive complex (*top line*). In the presence of mercury, MerR twists the DNA so as to properly align the promoter elements (*bottom line*).

GENERAL SUMMARY

We have described three mechanisms of activation:

1. ***Regulated Recruitment:*** CAP and λ repressor recruit an active polymerase (bearing a σ^{70} subunit) to one or another promoter.

2. ***Polymerase Activation:*** The activator NtrC works on an inactive enzyme (bearing a σ^{54} subunit) prebound to a promoter. The activator induces a conformational change in the enzyme, an effect that requires ATP.

3. ***Promoter Activation:*** MerR also works on a promoter bearing a prebound polymerase. In this case, the polymerase is inherently active (and bears σ^{70}), but the promoter is not. The activator induces a conformational change in the promoter, twisting it so that it resembles an active promoter—that is, one that allows polymerase to bind properly and initiate transcription.

These different modes of activation can be distinguished experimentally in a variety of ways, including the following.

- Activators that work by Mechanisms 1 and 2 bear DNA-binding and activating regions that can be independently inactivated by mutation. This is not the case for activators that work by Mechanism 3.
- An overexpressed activator lacking a DNA-binding function will activate only if it works as in Mechanism 2.
- For regulated recruitment, ordinary activating region-target interactions can be replaced by other protein-protein interactions; by additional protein-DNA interactions; or by increased polymerase concentration. That is, activator bypass experiments work. To the extent tested, this is not true for Mechanisms 2 and 3.
- Many different activator-polymerase interactions suffice for activation by regulated recruitment: there is no "privileged site" for contact on the enzyme. For Mechanism 2, there is a special site that, as far as we know, must be touched to activate the enzyme.
- For Mechanism 3, a highly specialized interaction with DNA is required. For the other cases, DNA binding serves merely a "locating function," and thus there are many different detailed modes of DNA binding of the activator that work.

As we turn to yeast, we will encounter many complications, but these models and experimental strategies will prove useful in identifying basic mechanisms of gene regulation.

MORE ON REPRESSION IN BACTERIA

We have noted that, especially where genes are controlled by regulated recruitment, specific repressors often work in opposition to activators. In the cases we discussed (Lac and λ), the repressors bind to sites overlapping promoters and exclude polymerase. It has also been proposed that repressors bound to sites flanking a promoter can interact to form a tight loop that excludes polymerase. But other ways to repress transcription are found in bacteria as well. These include: "silencing," where proteins bind over extended regions of DNA and exclude polymerase; inhibition of DNA-bound activators; and inhibition of DNA-bound polymerase. Here are examples of each of these alternative mechanisms.

Silencing

The *E. coli* protein called HNS binds DNA with no known sequence specificity. A gene called *bgl* is covered and "silenced" by this protein; what causes the protein to selectively bind there is not known. The gene bears a weak CAP site from which CAP is excluded by HNS. CAP will activate the gene if the cells are HNS-deficient or if the CAP site is mutated so that it binds CAP more tightly. Another way to activate the gene is to introduce binding sites for either Lac or λ repressor near the CAP site and to supply the corresponding repressor along with CAP. Presumably the repressors bind sufficiently tightly to their sites to disrupt HNS binding and thereby allow access of CAP to its weak site.

SopB is a membrane-bound protein that bears a specific DNA-binding domain exposed to the cytoplasm. Ordinarily, the protein binds certain plasmid DNA molecules to the membrane and thereby helps ensure their proper segregation at cell division. When overexpressed, the protein silences plasmid genes located within some 10 kb of its specific binding sites. The suggested mechanism is as follows. Overexpression of SopB results in a dense patch of exposed SopB DNA-binding domains at the membrane. A region of DNA brought to that patch by interaction of SopB with one or more specific SopB-binding sites would tend to bind, nonspecifically, to the other exposed SopB DNA-binding domains as well. This picture is supported by the finding that the membrane-binding domain of SopB is required for silencing and that the domain mediates silencing when attached to a heterologous DNA-binding domain, provided the DNA bears the corresponding binding site.

Inhibiting an Activator

Recall that a λ repressor bound at O_{R1} helps another dimer of repressor bind to the adjacent site O_{R2}. That repressor in turn contacts polymerase

(*Continued on following page.*)

and activates transcription from P_{RM}. In a construct in which O_{R1} is moved one or a few turns of the DNA helix upstream, activation is abolished. In this case, repressor still binds cooperatively to the two sites (with the intervening DNA forming a small loop). Evidently, interaction between repressor at O_{R2} with repressor at the upstream site subtly alters the way repressor sits at O_{R2} and eliminates its interaction with polymerase.

CytR is an example of a repressor that normally works by inhibiting an activator, in this case CAP. At the *cytR* promoter, it binds to a site adjacent to, and downstream from, the CAP site and covers the CAP-activating region.

Inhibiting Polymerase

The P4 protein, encoded by a phage that grows on the bacterium *Bacillus subtilus,* binds at the *P-A2c* promoter and prevents polymerase escape. The polymerase binds strongly to this promoter, and the additional interaction between the regulator and polymerase is sufficient to prevent escape of the polymerase from the promoter. The same regulator, making the same contact with polymerase, activates transcription from another, weaker, promoter.

These effects are explained by the fact that promoters differ in the ease with which polymerase escapes the open complex as transcription initiates. Where that step is difficult, an additional protein-protein interaction between activator and polymerase can prevent escape. At other promoters, escape occurs readily, and interaction with the regulator activates transcription by recruitment of polymerase.

Experiments with the Gal repressor of *E. coli* suggest that this protein too can repress by interacting with polymerase. In this case, it is believed that the inhibition is caused by the repressor interacting with polymerase in such a fashion as to prevent the transition from the closed to the open complex at the *gal* promoter. In contrast to the action of the P4 protein, this inhibitory effect of the Gal repressor is highly sensitive to the precise positioning of the repressor-binding site in relation to the promoter.

FOOTNOTES

[1]Another protein, called ω, is associated with RNA polymerase. ω holds together the amino and carboxyl termini of β', and thereby helps that protein fold. Once the folded β' has been incorporated into the polymerase, ω can be removed without affecting enzyme activity in vitro. In vivo, elimination of ω slows down formation of active polymerase but is not lethal.

The two σ factors considered in the text (σ^{70} and σ^{54}) are made constitutively. Some σ factors are induced by environmental signals. For example, heat shock induces expression of a form of σ that directs the enzyme to a set of genes, expression of which helps to overcome the shock. Changes of σ factors can drive developmental programs. For example, the bacterium *Bacillus subtilus* undergoes a series of changes to form spores, and each stage is characterized by the synthesis of a new σ factor that directs the enzyme to a new set of genes.

[2]The operator sequence is approximately twofold rotationally symmetric, and the center of that symmetry lies under that of the two binding domains of the protein. Parts of the protein in addition to the HTH motif can contribute to both the energy and specificity of the binding. Proteins such as the Lac repressor that bind as tetramers or even high-order oligomers typically contact an operator using a pair of subunits as shown in Figure 1.4. The orientation of the HTH motif is inverted in the Lac repressor compared with its orientation in CAP and other DNA-binding proteins we will encounter, e.g., λ repressor and Cro. Most DNA-protein recognition is effected using a "recognition α-helix," but other ways, including use of a β-sheet, are known.

[3]In the Lac repressor, the HTH motifs are presented on small domains, called the headpieces, attached to the main body of the protein. In the absence of inducer, the headpieces are rigidly held apart at the same distance that separates successive segments of the major groove along one face of the DNA. In the induced conformation, however, the headpieces are free to wiggle independently of each other. The added entropic cost of "fixing" the headpieces greatly reduces specific binding.

[4]The phrase "increasing local concentration," a useful shorthand, can be misleading. Merely increasing the local concentration of a protein at a given site by introducing a second site for that same protein nearby will not, per se, increase the occupancy of the first site. Rather, the effect requires a simultaneous interaction between the proteins bound to the two sites. Likewise, if there is an interaction between the two proteins, but that interaction covers the DNA-binding surface of one of them, again the local concentration might be increased, but DNA binding will not be.

[5]If site B in Figure 1.6 were modified so as to bind its protein considerably more tightly than site A binds protein A, binding of A to its site would depend on the presence of B, but binding of B might not require A. The discussion here and in the text also applies if A and B are two identical proteins that interact with each other while binding separate sites. More than two proteins can bind cooperatively to DNA. All of these variations are found in nature.

[6]CAP has three activating regions. The two others touch the main body of α and σ, respectively. These additional activating regions have no role at the *lac* genes, but they do at other genes where the CAP-binding site is found at different positions relative to the polymerase-binding site.

[7] One might imagine two ways that transcription would proceed once polymerase was tethered to DNA as in these experiments. In the first, the polymerase would maintain contact with the upstream tethering site and the DNA would be threaded past the immobilized polymerase. In the second, the upstream polymerase-DNA contacts would be broken as the polymerase moved away from the promoter. The latter is evidently the rule: a variety of experiments indicate that if the upstream protein-DNA interaction becomes too strong, that interaction can repress rather than activate transcription (see panel on More on Repression in Bacteria).

[8] The following experiment emphasizes the simple adhesive roles of the –35 and UP-elements. A truncated form of σ^{70}, lacking its –35 region recognition domain, was generated. Polymerase bearing the truncated σ does not work at a promoter bearing the ordinary –10 and –35 sequences. It does work, however, if a CAP site is introduced and active CAP is present. The CAP-α interaction compensates for the loss of the interaction between σ and the –35 region on DNA.

[9] It has been reported that the Lac repressor and polymerase form a stable tripartite complex with DNA at the *lac* promoter. But other studies indicate that formation of this complex occurs only under nonphysiological conditions in vitro. And in vivo (as measured in footprinting experiments), polymerase is not found at the promoter in conjunction with repressor. So we have adopted the view that repressor works by excluding binding of RNA polymerase, rather than by forming a stable inactive complex.

[10] Later in the λ life cycle, the concentration of Cro reaches a point at which it binds to O_{R1} and O_{R2} and turns down transcription of lytic phage genes as well.

[11] We say "stimulates cleavage" because RecA does not bear a typical protease active site. Rather, it holds the repressor in a conformation that stimulates an auto-cleavage reaction.

[12] Both of these protein-protein interactions are sufficiently weak that, at the concentration of repressor found in the cell, a significant fraction of the unbound protein is present as monomer.

[13] If the λ system worked as it does at *lac*, where the repressor is a stable tetramer whether interacting with DNA or not, then a much larger fraction of repressor molecules would need to be inactivated to achieve a similar level of induction.

[14] Repressor at O_{R2} can bind cooperatively *either* with another dimer at O_{R1} *or* with another dimer at O_{R3}. The phenomenon is called "alternate pairwise cooperativity." Because O_{R1} binds repressor significantly more tightly than does O_{R3}, the former reaction predominates.

[15] Only recently has it been demonstrated that autogenous negative control by repressor, in a lysogen, is physiologically important for allowing efficient induction. The degree of self-repression was previously underestimated for the following reason, one that raises a common methodological problem. As described in *A Genetic Switch*, the workings of the switch were deciphered (in part) by attaching regulatory elements (e.g., O_R), to reporter genes (e.g., the *lacZ* gene), and supplying the regulators (e.g., λ repressor) from independently manipulated promoters. This "reductionist" approach could not reveal the effects of proteins binding to distal

elements—in this case repressor binding to O_L—and so the degree of autogenous self-regulation was significantly underestimated.

[16]Consistent with these ideas are the two findings alluded to above, namely, that CAP can work at P_{RM} and that activator bypass experiments work at both the *lac* and λ promoters. In addition, a single-amino-acid change in polymerase alters the effect of λ repressor at P_{RM} from one on isomerization to one (largely) on initial polymerase binding.

[17]Were both activators to contact the polymerase simultaneously, the total energy would be the sum of the two interactions considered separately. Because, in a simple Mass Action formulation, the energies of interaction contribute exponentially to the binding constant, the effect on recruitment would be synergistic, as defined here.

[18]To our knowledge, activator bypass experiments, other than providing high concentrations of polymerase in vitro, have not been attempted for this case.

[19]All of these activators typically work when bound well upstream of their target genes. Some genes bear separate binding sites for two such activators, and those genes can be controlled by either activator. For example, certain genes activated by NtrC are also activated—independently—by the related activator NifA, which senses oxygen and ammonia levels.

BIBLIOGRAPHY

Books

Baumberg S., ed. 1999. *Prokaryotic gene expression.* Oxford University Press, Oxford, United Kingdom.

Cold Spring Harbor Symposia on Quantitative Biology. 1998. Volume 63: *Mechanisms of transcription.* Cold Spring Harbor Laboratory Press, Cold Spring Harbor, New York.

Echols H. 2001. *Operators and promoters: The story of molecular biology and its creators* (ed. C.A. Gross). University of California Press, Berkeley.

Müller-Hill B. 1996. *The lac operon.* de Gruyter, Berlin.

Ptashne M. 1992. *A genetic switch: Phage λ and higher organisms,* 2nd edition. Blackwell Science, Malden, Massachusetts and Cell Press, Cambridge, Massachusetts.

RNA Polymerase, Promoters, and DNA Binding

Barrios H., Valderrama B., and Morett E. 1999. Compilation and analysis of σ(54)-dependent promoter sequences. *Nucleic Acids Res.* **27:** 4305–4313.

Bell C.E. and Lewis M. 2001. The Lac repressor: A second generation of structural and functional studies. *Curr. Opin. Struct. Biol.* **11:** 19–25.

Burgess R.R. and Anthony L. 2001. How sigma docks to RNA polymerase and what sigma does. *Curr. Opin. Microbiol.* **4:** 126–131.

Busby S. and Ebright R.H. 1994. Promoter structure, promoter recognition, and transcription activation in prokaryotes. *Cell* **79:** 743–746.

Darst S.A. 2001. Bacterial RNA polymerase. *Curr. Opin. Struct. Biol.* **11:** 155–162.

Ebright R.H. 2000. RNA polymerase: Structural similarities between bacterial RNA polymerase and eukaryotic RNA polymerase II. *J. Mol. Biol.* **304:** 687–698.

Gourse R.L., Ross W., and Gaal T. 2000. UPs and downs in bacterial transcription initiation: The role of the alpha subunit of RNA polymerase in promoter recognition. *Mol. Microbiol.* **37:** 687–695.

Gross C.A., Chan C., Dombroski A., Gruber T., Sharp M., Tupy J., and Young B. 1998. The functional and regulatory roles of sigma factors in transcription. *Cold Spring Harbor Symp. Quant. Biol.* **63:** 141–155.

Hawley D.K. and McClure W.R. 1983. Compilation and analysis of *Escherichia coli* promoter DNA sequences. *Nucleic Acids Res.* **11:** 2237–2255.

Luscombe N.M., Austin S.E., Berman H.M., and Thornton J.M. 2000. An overview of the structures of protein-DNA complexes. *Genome Biol.* **1:** REVIEWS001.

McClure W.R. 1985. Mechanism and control of transcription initiation in prokaryotes. *Annu. Rev. Biochem.* **54:** 171–204.

Activation and Repression

Adhya S., Geanacopoulos M., Lewis D.E., Roy S., and Aki T. 1998. Transcription regulation by repressosome and by RNA polymerase contact. *Cold Spring Harbor Symp. Quant. Biol.* **63:** 1–9.

Atlung T. and Ingmer H. 1997. H-NS: A modulator of environmentally regulated gene expression. *Mol. Microbiol.* **24:** 7–17.

Buck M., Gallegos M.T., Studholme D.J., Guo Y., and Gralla J.D. 2000. The bacterial enhancer-dependent sigma(54) (sigma(N)) transcription factor. *J. Bacteriol.* **182:** 4129–4136.

Busby S. and Ebright R.H. 1999. Transcription activation by catabolite activator protein (CAP). *J. Mol. Biol.* **293:** 199–213.

Choy H. and Adhya S. 1996. Negative control. In Escherichia coli *and* Salmonella: *Cellular and molecular biology,* 2nd edition (ed. F. Neidhardt et al.), pp. 1287–1299. ASM Press, Washington, D.C.

Geiduschek E.P., Fu T.-J., Kassavetis G.A., Sanders G.M., and Tinker-Kulberg R.L. 1997. Transcriptional activation by a topologically linkable protein: Forging a connection between replication and gene activity. *Nucleic Acids Mol. Biol.* **11:** 135–150.

Hochschild A. and Dove S.L. 1998. Protein-protein contacts that activate and repress prokaryotic transcription. *Cell* **92:** 597–600.

Lloyd G., Landini P., and Busby S. 2001. Activation and repression of transcription initiation in bacteria. *Essays Biochem.* **37:** 17–31.

Magasanik B. 2000. Global regulation of gene expression. *Proc. Natl. Acad. Sci.* **97:** 14044–14045.

Müller-Hill B. 1998. Some repressors of bacterial transcription. *Curr. Opin. Microbiol.* **1:** 145–151.

Ptashne M. and Gann A. 1997. Transcriptional activation by recruitment. *Nature* **386:** 569–577.

Rine J. 1999. On the mechanism of silencing in *Escherichia coli. Proc. Natl. Acad. Sci.* **96:** 8309–8311.

Rojo F. 2001. Mechanisms of transcriptional repression. *Curr. Opin. Microbiol.* **4:** 145–151.

Rojo F., Mencia M., Monsalve M., and Salas M. 1998. Transcription activation and repression by interaction of a regulator with the alpha subunit of RNA polymerase: The model of phage phi 29 protein p4. *Prog. Nucleic Acid Res. Mol. Biol.* **60:** 29–46.

Rombel I., North A., Hwang I., Wyman C., and Kustu S. 1998. The bacterial enhancer-binding protein NtrC as a molecular machine. *Cold Spring Harbor Symp. Quant. Biol.* **63:** 157–166.

Roy S., Garges S., and Adhya S. 1998. Activation and repression of transcription by differential contact: Two sides of a coin. *J. Biol. Chem.* **273:** 14059–14062.

Schleif R. 2000. Regulation of the L-arabinose operon of *Escherichia coli. Trends Genet.* **16:** 559–565. (Interesting case not discussed in the text.)

Shapiro L. and Losick R. 1997. Protein localization and cell fate in bacteria. *Science* **276:** 712–718. (Interesting cases not discussed in the text.)

Valentin-Hansen P., Sogaard-Andersen L., and Pedersen H. 1996. A flexible partnership: The CytR anti-activator and the cAMP-CRP activator protein, comrades in transcription control. *Mol. Microbiol.* **20:** 461–466.

Xu H. and Hoover T.R. 2001. Transcriptional regulation at a distance in bacteria. *Curr. Opin. Microbiol.* **4:** 138–144.

Yarmolinsky M. 2000. Transcriptional silencing in bacteria. *Curr. Opin. Microbiol.* **3:** 138–143.

Antitermination

Friedman D.I. and Court D.L. 1995. Transcription antitermination: The λ paradigm updated. *Mol. Microbiol.* **18:** 191–200.

Greenblatt J., Mah T.F., Legault P., Mogridge J., Li J., and Kay L.E. 1998. Structure and mechanism in transcriptional antitermination by the bacteriophage λ N protein. *Cold Spring Harbor Symp. Quant. Biol.* **63:** 327–336.

Roberts J.W., Yarnell W., Bartlett E., Guo J., Marr M., Ko D.C., Sun H., and Roberts C.W. 1998. Antitermination by bacteriophage λ Q protein. *Cold Spring Harbor Symp. Quant. Biol.* **63:** 319–325.

Yanofsky C. 2000. Transcription attenuation: Once viewed as a novel regulatory strategy. *J. Bacteriol.* **182:** 1–8. (Interesting case not discussed in the text.)

CHAPTER 2

Yeast: A Single-celled Eukaryote

Influential ideas are always simple. Since natural phenomena need not be simple, we master them, if at all, by formulating simple ideas and exploring their limitations.

AL HERSHEY

THE INSIDE OF A YEAST CELL LOOKS MORE LIKE THAT OF A HUMAN CELL than that of a bacterium. The DNA is wrapped around proteins called histones to form bead-like structures called nucleosomes, and the chromosomes are sequestered in a cellular compartment called the nucleus. For these and other reasons, yeast is classified as a eukaryote, as are humans, flies, worms, and plants.

Expression of a typical eukaryotic gene is a more complicated undertaking than is expression of a bacterial gene. As we noted in the Introduction, eukaryotic RNA must be spliced, modified in various ways, and transported out of the nucleus to the cytoplasm where it can be translated into protein. We will mostly ignore these complications and will continue to focus on the initiation of transcription of a gene and, to a limited extent, control of elongation of the RNA once transcription has been initiated. But even these processes are complicated by nucleosomes and nuclei. As discussed below, eukaryotes have enzymes that modify nucleosomes, modifications that can affect protein binding to DNA. And the sequestration of genes in the nucleus means that in order to perform their task, regulators often must move from one compartment (the cytoplasm) to another (the nucleus).

Much of what we know about eukaryotic gene regulation comes from studies of the yeast *Saccharomyces cerevisiae*. This organism grows rapidly, about 20-fold faster than mammalian cells and only some 3-fold more slowly than *Escherichia coli*. It is unicellular and can be propagated as a haploid or a diploid. Mutants can be selected or recognized by simple assays, and sequences in and around genes can be altered at will. The genome, completely sequenced, comprises about 6000 genes, only about 2000 more than *E. coli*. Unless stated otherwise, we use the term yeast to refer to *S. cerevisiae*.

Our main intent in this chapter is to ask which of the three models for bacterial gene activation best describes how activators work in yeast. We will find that, despite many added complexities and some uncertainties, these eukaryotic activators work by recruitment.

RNA POLYMERASE

Most yeast genes are transcribed by the enzyme RNA polymerase II (PolII). This enzyme bears striking structural similarities to the bacterial enzyme. The stable core polymerase is readily isolated from cells and, like its bacterial counterpart, transcribes DNA nonspecifically.[1]

OTHER PARTS OF THE TRANSCRIPTIONAL MACHINERY

In yeast, as in bacteria, activators and repressors regulate gene expression. But the transcriptional machinery itself is vastly more complex: at least 50 proteins, in addition to the core polymerase, can be involved in transcribing a gene. Table 2.1 presents a list of many of these auxiliary proteins, according to their standard classifications, and with an explanation of what some of the abbreviated names stand for. We do not know the functions of many of these proteins, but can nevertheless make the following general points:

- Some of these proteins bind DNA directly and form a platform for polymerase. These include TBP (TATA-binding protein), TFIIB, TAF17, and TFIIA.
- Some of the proteins are enzymes. TFIIH, for example, contains both a kinase and two helicases (enzymes that unwind DNA), and it helps polymerase initiate transcription and then escape the promoter. Other proteins help or hinder transcription by chemically modifying histones in nucleosomes. These include histone acetylases (HATs) and histone de-

TABLE 2.1. Transcriptional machinery: A simplified view

Transcriptional machinery	Components	No. subunits
RNA polymerase II	Rpb1-12	12
General transcription factors (GTFs)	TBP	1
	TFIIA	2
	TFIIB	1
	TFIIE	2
	TFIIF	3
	TFIIH	9
TBP-associated factors (TAFs)	TAF145, TAF17, etc.	11
Mediator	Srb2,4–11	26
	Med1,2,4,6,7,8,11	
	Gal11 and others	
Others	Elongation complex	
	NC2	
Nucleosome modifiers		
Histone acetylases (HATs)		
SAGA	**Gcn5;** Spt3,7,8,20; Ada1,2,3; TAFs; Tra1	15
Ada	**Gcn5;** Ada2,3; Ahc1	4
NuA3	**Sas3;** TAFs, etc.	
NuA4	**Esa1;** Tra1, etc.	
Nucleosome remodellers		
Swi/Snf	**Snf2**,4,6,9; Swi1,2,3	11
RSC	**Sth1;** Rsc1-10	17
Histone deacetylases (HDACs)		
	Rpd3	
	Hda1	
	Hos1,2,3	
	Sir2	

TBP binds to DNA sequences resembling TATA found at the beginning of many genes (hence, the name, TATA-binding protein). TBP plus associated TAFs is called TFIID. TAF145 binds TBP directly and bears a HAT activity. The mediator can be dissociated into subcomplexes, including one containing Gal11 and Med2, and another containing Srb8-11. Some of the subunits of TFIIH are required not for transcription but rather for repair of damaged DNA. Components shown in bold indicate the catalytic subunit of nucleosome modifiers. The number of subunits is given for some complexes; many are estimates. The names of various proteins and complexes are based on techniques used in their identification. For example, the TFIIs (transcription factors involved in transcription by PolII) are different biochemical fractions (B, E, etc.) that are required for transcription in vitro. The names of some of the components of mediator and nucleosome modifiers are based on one or another genetic screen that identified them.

acetylases (HDACs), and histone kinases, for example. Protein complexes called remodelers apparently alter the disposition of nucleosomes without modifying them covalently.[2]

- Some of these proteins are globally required—when any one of them is depleted from cells, transcription of all genes by PolII ceases. These include TBP, TFIIB, and the helicases in TFIIH. In contrast, other proteins are required for full expression of only certain genes under certain conditions. These include the nucleosome modifying enzymes and remodelers, and several components of the so-called "mediator" complex (see below).
- Most proteins of the transcription machinery can be isolated from cells as parts of complexes. For example, the mediator complex includes more than 20 proteins. That complex, in turn, can associate with polymerase to form a "holoenzyme." Thus, despite the large number of proteins involved in transcription, it is likely that the number of preformed complexes that must assemble at a gene for transcription is not large, perhaps only a few. [3]

Two general features of these complexes are worth noting:

- Proteins found in one complex are sometimes found in other complexes as well. For example, certain TBP-associated factors (TAFs) associate not only with TBP in the TFIID complex, but also with the HAT Gcn5 in the SAGA (*S*pt, *A*da, *G*cn5 *A*cetyltransferase) complex. Gcn5 itself is part of at least one additional complex called Ada.
- At least some complexes are not formed by highly interdependent interactions. For example, deletion of Gcn5 evidently leaves the remainder of SAGA intact, and that diminished complex contributes to gene activation in certain cases.

Please recall our goal: we aim not so much to understand the complexities of transcription per se, but rather to understand how that process is regulated—i.e., how the decision is made to transcribe one rather than another gene. The overview of the transcriptional machinery (including nucleosome modifiers) provided here is far from exhaustive. But even many of these details can be set aside as we analyze the general mechanism of activation.

AN OVERVIEW OF ACTIVATION

In Chapter 1 we described three modes of gene activation in bacteria. The diagnostic tests summarized at the end of that chapter (see page 49) distinguish between different mechanisms in bacteria. Those same tests can be applied to yeast activators as well. The results of such tests are consistent with predictions of the "regulated recruitment" mechanism and are inconsistent with those of the other bacterial models. Among salient findings for the typical gene, we will see the following.

- The transcriptional machinery is not bound to the promoter prior to activation.
- Yeast activators have separable activating and DNA-binding domains.
- An activating region must be tethered to DNA near a promoter to activate transcription.
- A given activating region will work when attached to a heterologous DNA-binding domain, even one taken from bacteria.
- Any of a wide array of genes can be brought under control of a given activator by introducing the appropriate activator binding sites nearby.
- Various "activator bypass" experiments work.
- Many different peptide sequences work as activating regions; their properties are more suggestive of adhesive surfaces than of specialized functions.

This analysis will reveal the general mechanism by which activators work. We will then be in a position to discuss how recruitment of polymerase itself is effected in the face of the many complexities of the transcriptional machinery.

A MODEL CASE: THE *GAL* GENES

We concentrate on one set of genes, those required for metabolism of the sugar galactose. Just as for the *gal* genes of *E. coli*, transcription is induced by galactose.

The degree of induction of the *GAL* genes is unusually large for yeast genes, a useful feature. Cells growing in the absence of galactose express these genes at very low levels, and addition of galactose induces them more than 1000-fold. When induced, the *GAL* genes are among the most highly expressed in the cell. This high level of expression is due to the action of a single regulator, the activator Gal4, which binds sites upstream of the *GAL* genes. Gal4 will also activate any of a wide array of yeast genes if Gal4-binding sites are introduced nearby. And, as we shall see in the next chapter, Gal4 can work in many higher eukaryotes as well.

A good deal of this chapter is devoted to analysis of how Gal4 works. Some of the experiments we will describe were first performed with other yeast activators, but the important points can be illustrated using Gal4. There is one additional regulator of the *GAL* genes: the repressor Mig1. This repressor works only in the presence of glucose, helping to ensure that if both galactose and glucose are present, glucose will be utilized preferentially.

We begin with a brief overview of the activities of Gal4 and Mig1, and then, keeping in mind the three bacterial models of the last chapter, we delve into the question of how Gal4 works. After this analysis, we return to the action of Mig1.

Overview of Regulation of a *GAL* Gene

Figure 2.1 shows one of the *GAL* genes, *GAL1*. The four sites to which Gal4 binds comprise the so-called galactose upstream activating sequence (UAS_G). This element lies about 275 bp upstream of the promoter. A binding site for the repressor Mig1 lies between the UAS_G and the transcriptional start site.[4]

Yeast promoters are less well defined than their bacterial counterparts. Each *GAL* gene, like many other eukaryotic genes, bears a sequence called the TATA element. TBP (see Table 2.1) binds this sequence along with many other proteins including polymerase. The polymerase initiates transcription at a site about 85 bp downstream from TATA. Figure 2.2 shows three states of the *GAL1* gene.

- In the absence of galactose and glucose, Gal4 is bound to the UAS_G, but it does not activate transcription because it is complexed with the inhibitor Gal80 (Figure 2.2a).

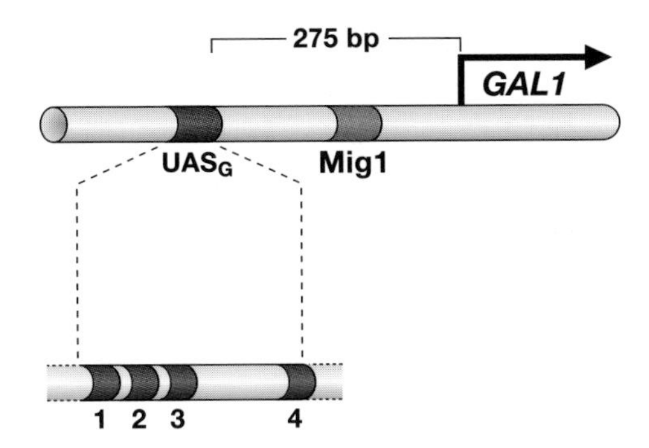

FIGURE 2.1. The *GAL1* gene and the region just upstream. The four binding sites for Gal4 within the UAS_G are indicated on the blown-up section. The Mig1-binding sites lie between the UAS_G and the gene. The UAS_G spans some 118 bp, each Gal4 site comprising 17 bp. (Modified, with permission, from Ptashne 1992 [see Chapter 1].)

- In the presence of galactose, Gal4 is freed from the inhibiting effect of Gal80, and the gene is transcribed (Figure 2.2b).
- Galactose and glucose are both present in Figure 2.2c, but the gene is kept off by the repressor Mig1. Mig1 does not work alone: it recruits a "repressing complex," a key component of which is a protein called Tup1. A Mig1 site is also found upstream of the *GAL4* gene, and so glucose reduces the level of Gal4 in addition to repressing *GAL1* directly. Mig1 sites are also found in front of many other yeast genes repressed by glucose.

Before considering how Gal4 works, we comment briefly on the mechanisms of DNA binding and signal detection by the two regulators of *GAL1*.

Specific DNA Binding

Gal4 dimers recognize each of the 17-bp sites in the UAS_G using a domain called a zinc cluster (see Figure 2.3). Like the helix-turn-helix (HTH) we encountered in bacteria, this motif inserts an α-helix (a recognition helix) into the major groove of DNA. Mig1 binds DNA using a related motif called a zinc finger. Yeast transcriptional regulators use other DNA-binding domains as well, including the homeodomain, which bears a structure related to the HTH motif.

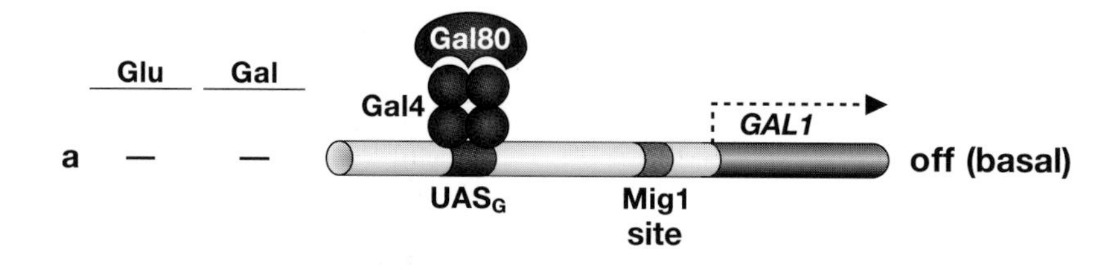

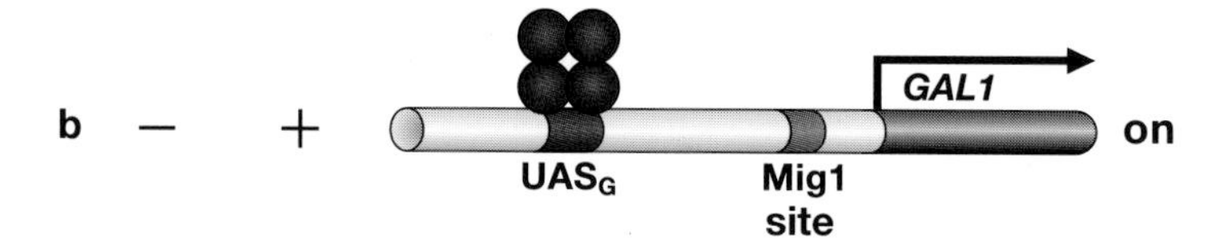

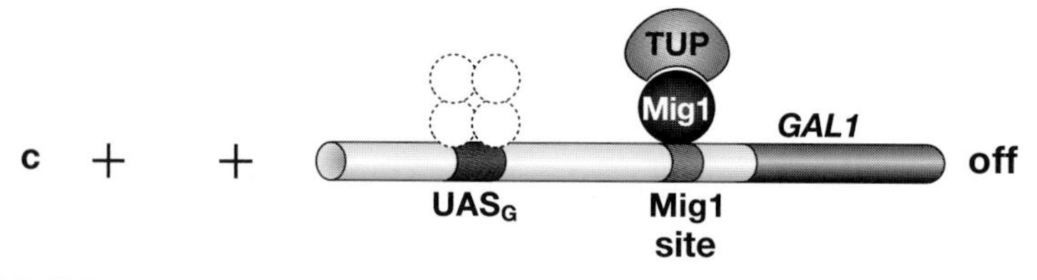

FIGURE 2.2. Three states of the *GAL1* gene. (*a*) Cells are grown in the "neutral" sugar raffinose, which has no effect on the *GAL* genes or their regulators. For simplicity, only a single Gal4 dimer is shown bound to the UAS_G. Gal4 bound to the UAS_G in the presence of glucose (*c*) is shown in dotted lines because, in addition to its direct effect on *GAL1*, Mig1 also represses the gene encoding Gal4. In *a*, the expression level is called "basal" to distinguish it from the fully repressed case (*c*), but that basal level is very low. Mig1 recruits the Tup1 complex (which also includes proteins not shown, Ssn6 being an example).

Detecting Physiological Signals

A protein called Gal3 senses galactose and conveys that signal to the gene: upon binding the sugar and ATP, Gal3 binds Gal80, an interaction that frees Gal4's activating region. This effect, it is believed, does not require dissociation of Gal80 from Gal4 but rather, evidently, some reorientation of Gal80 on Gal4.[5,6]

In the absence of glucose, Mig1 is held in the cytoplasm in a phosphorylated form. Glucose inhibits the kinase responsible for that phosphoryla-

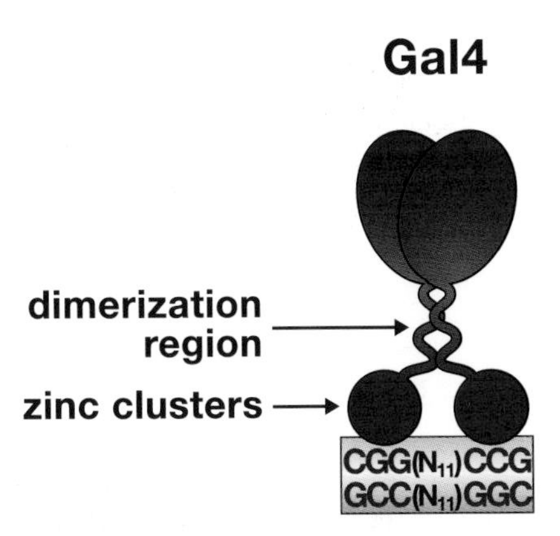

FIGURE 2.3. A dimer of Gal4 bound to a 17-bp site. Each of the zinc clusters recognizes the base-pair triplet indicated on the figure, and the triplets are separated by 11 bp. Activators related to Gal4 bear identical zinc clusters, but the spacing between the triplets is unique in each case. The formal name for these zinc clusters is Zn(II)Cys6 binuclear cluster. Such domains are not found in bacteria.

tion and thereby allows Mig1 to enter and remain in the nucleus. There it binds DNA and recruits a repressing complex as we have mentioned.[7]

The following generalization seems to hold widely for eukaryotes: in the absence of their respective inducing signal, transcriptional regulators tend not to be found in the nucleus with (in the case of activators) their activating regions free to work. Rather, activating regions are masked (as for Gal4), or, as with Mig1, the regulators are maintained outside the nucleus, until the inducing signal is detected. We comment on the significance of this generalization below (see Squelching).

HOW Gal4 WORKS

Separate DNA-binding and Activating Regions of Gal4

We have seen that certain bacterial activators bear two surfaces: one that contacts DNA and the other, the activating region, that interacts with RNA polymerase. In those cases, the *mode* of DNA binding is not critical: DNA binding serves merely to locate the activator near the gene to be activated. This characterization holds both for activators that work by regulated recruitment (e.g., CAP and λ repressor) and for activators that induce a

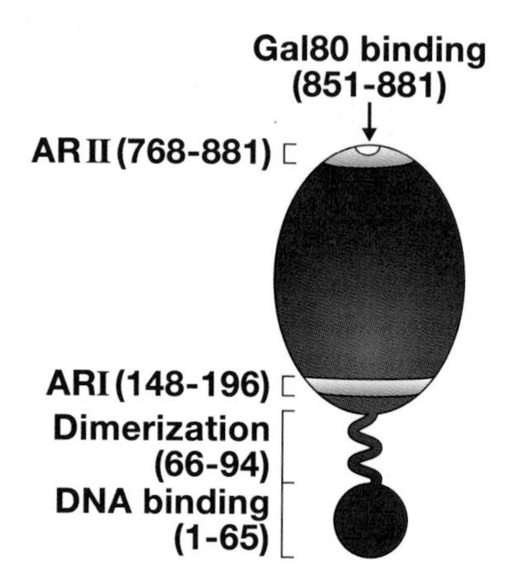

FIGURE 2.4. A Gal4 monomer. Gal4(1–100) and Gal4(1–147) form stable dimers but do not activate; they are widely used to construct fusion proteins that tether heterologous proteins to DNA. Note that there are two activating regions in Gal4, but AR (activating region) I has little effect in the intact molecule. The Gal80-binding site overlaps ARII. A nuclear localization sequence (NLS) is found within residues 1–74.

conformational change in a polymerase bound to a nearby promoter (e.g., NtrC). In contrast, for activators that work like the bacterial activator MerR, the specific mode of DNA binding is inextricably intertwined with activation. In these cases, the activating event is a change in the conformation of DNA to which the polymerase is bound.

The following experiments show that Gal4 has separate DNA-binding and activating functions. These functions are carried on separate domains of the protein, and the specific mode of DNA binding is not critical for activation. A fragment of Gal4, comprising the first 100 of its 881 amino acids, contains the DNA-binding region but lacks any activating region. This fragment, designated Gal4(1–100), also contains a sequence that directs the protein to the nucleus, and another sequence that mediates dimerization (see Figure 2.4).

Gal4(1–100) binds DNA but does not activate transcription (because it lacks the activation function). It is therefore analogous to the *pc* mutants of activators discussed in Chapter 1. The complementary fragment of

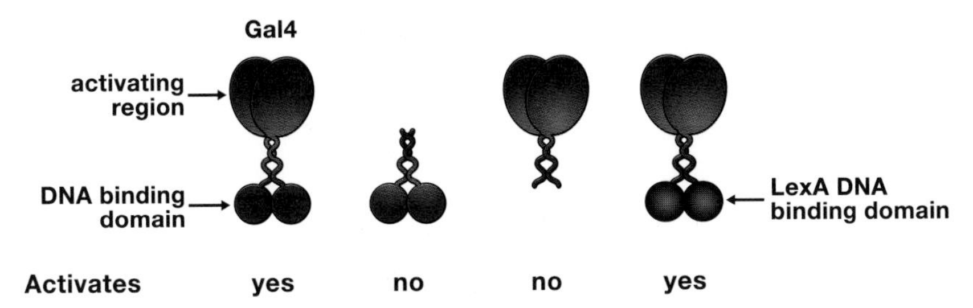

FIGURE 2.5. A domain swap. Gal4's activating region has been attached to many different DNA-binding domains, and, conversely, many different activating regions have been attached to the Gal4 DNA-binding domain. In each case, the hybrid protein activates transcription of a gene bearing the appropriate DNA-binding sites. LexA presumably lacks an NLS, and we do not know if the hybrid enters the nucleus passively or if the Gal4 fragment has an NLS in addition to that found in its first 74 residues.

Gal4 (residues 100–881) can be attached to a different DNA-binding domain in an experiment called a "domain swap." The hybrid protein activates a gene bearing binding sites recognized by the new DNA-binding domain.

Figure 2.5 shows a domain swap experiment between Gal4 and the bacterial repressor LexA. This hybrid, designated LexA-Gal4, activates a gene bearing LexA sites, whereas LexA alone does not. The hybrid activator—like Gal4 itself—works when positioned in front of any of a wide array of genes.[8]

These experiments reveal three important features of Gal4:

- DNA binding does not suffice for activation—an activating region is also required.
- The activating region must be tethered to DNA to work.
- The same activating region will work when attached to any of a variety of DNA-binding domains, even one from bacteria.

A surmise from this third point is that there are few, if any, stereospecific restrictions on how the activating region must be attached to the DNA-binding domain. The experiment of Figure 2.6 further demonstrates this point by showing that the link between the activating region and the DNA-binding domain need not be covalent. In the depicted experiment,

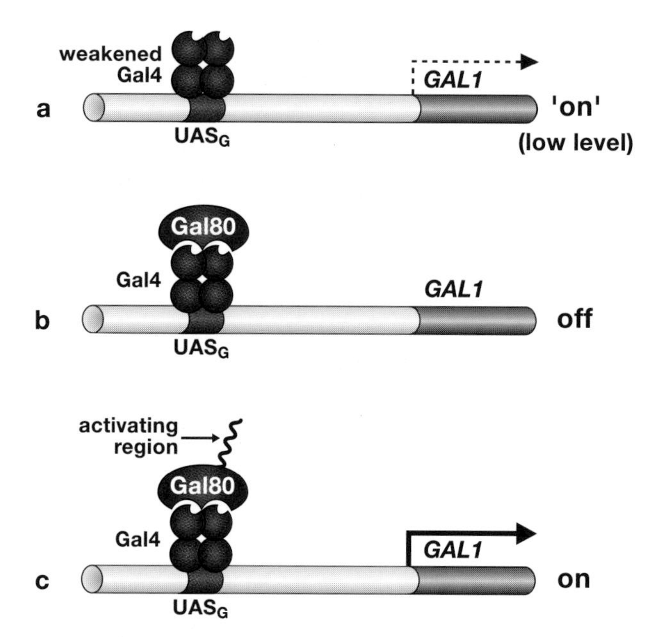

FIGURE 2.6. Creating a two-component activator. The Gal4 derivative used in this experiment bears a weakened activating region, but nevertheless binds Gal80. Addition of Gal80 bearing an attached strong activating region results in an increase in transcription.

an activating region was attached to Gal80, the protein that binds to and inhibits Gal4. This Gal80 derivative, however, activates transcription upon binding to DNA-bound Gal4. The Gal4 DNA-binding domain in effect tethers the Gal80 derivative to DNA so that it activates transcription of the nearby gene.

An implication of this result is that any protein-protein interaction that brings an activating region to DNA will elicit transcription. This is the basis of the "two-hybrid" system that detects pairs of interacting proteins as illustrated in Figure 2.7.

These observations show that Gal4 does not work like the bacterial regulator MerR. In that case, as we reviewed above, the mode of DNA binding was crucial to activation. The following sections clarify which of the remaining two bacterial models best describes the action of Gal4. We begin with a closer look at Gal4's activating regions (there are two, see Figure 2.4) and then describe how they work.

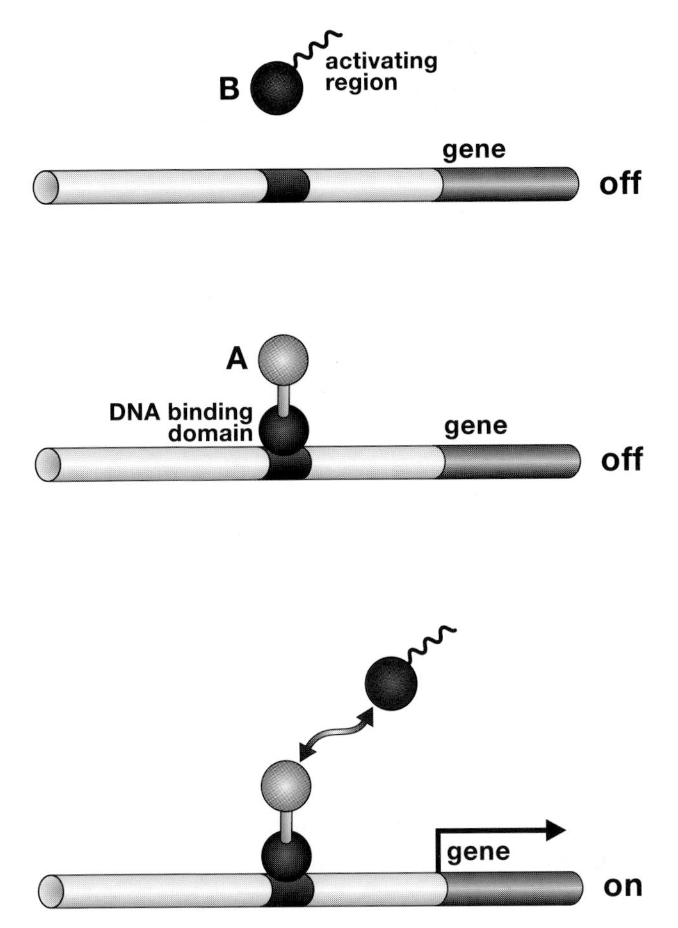

FIGURE 2.7. The two-hybrid assay. Two hybrid proteins are involved here: one comprises a DNA-binding domain attached to protein A; the other comprises an activating region attached to protein B. Neither hybrid protein activates transcription in the absence of the other. Interaction of A with B tethers the activating region to DNA, and the gene is activated.

Activating Region Structure

The two activating regions of Gal4, like those of many other yeast activators, are "acidic": they bear an excess of negatively charged residues (glutamic acid and aspartic acid), in addition to hydrophobic residues. These activating regions have properties that suggest they are adhesive-like surfaces, as revealed by the following experiments.[9]

TABLE 2.2. Improving an activator by mutation

Protein		Activating region charge	β-galactosidase activity
Gal4(1–147) + ARI	wild type	–7	710
	mutant 1	–8	970
	mutant 2	–9	1628
	mutant 3	–10	2484
Gal4(1–881)	wild type		3320

Mutant 2 bears two independently isolated mutations brought together in a single activating region. Mutant 3 carries yet an additional mutation isolated independently and added to the first two. The various activators were assayed using a reporter gene in which the *GAL1* promoter and its associated UAS_G were fused to the bacterial *lacZ* gene. The level of β-galactosidase produced in the yeast is taken as a measure of the strength of activation. Although not always indicated, such *GAL1-lacZ* reporters are used to measure activation in many of the experiments described in this chapter. (Modified, with permission, from Ptashne 1992 [see Chapter 1].)

Independent Mutations Can Increase Activator Strength Additively When Combined

Recall from Figure 2.4 that Gal4 has two activating regions, ARI (activating region I) and ARII. Each is about 100 amino acids long, and fusing either region to a DNA-binding domain creates an activator. Of the two fusions, the one bearing ARI activates more weakly than the other. Mutants of the weaker activator that activate more efficiently can be isolated. The improved activating regions are more acidic than the parent. The mutations usually eliminate one or more of the few positively charged residues—lysine and arginine. When two such mutations are combined, the doubly mutant activator works even better than either single mutant. Introduction of further mutations, by recombination, creates activators that work better still (Table 2.2).

Activating Regions Work with an Efficiency Proportional to Their Lengths

Among a set of deletion derivatives of the activating region ARII, the strength of activation is approximately proportional to the length of the remaining activating region fragment (Figure 2.8). Moreover, whereas replacing single hydrophobic residues in the full-length activating region has little effect, the same change in shorter (~15 residues) derivatives has a dramatic damaging effect.

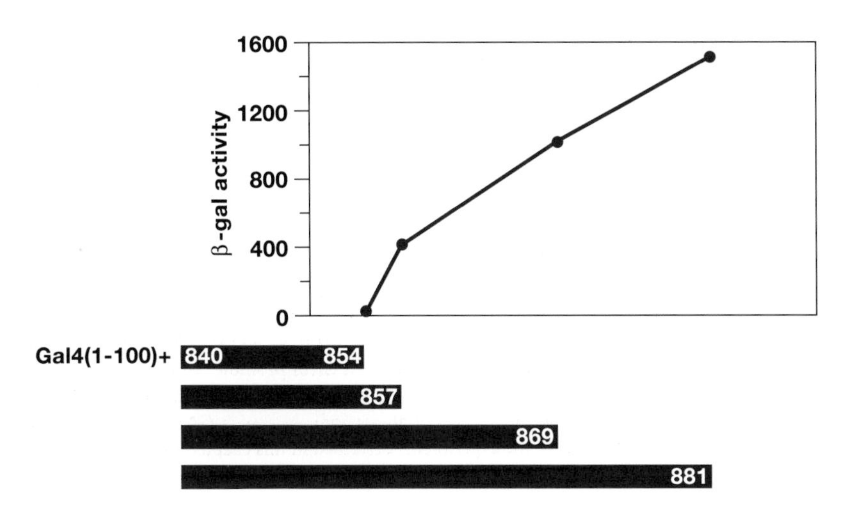

FIGURE 2.8. Activation proportional to activating region length. The red bars delineate the length of the activating region fragment attached to the Gal4 DNA-binding domain.

A plausible surmise from these (and other) experiments is that an activating region comprises reiterated binding sites for its target(s). Accordingly, a single mutation in a longer activating region (i.e., one with more activating elements) would be less sensitive to a single-amino-acid change than would be a shorter activating region.[10]

Activating Region–Target Interactions Tolerate Sequence Changes

ARII, one of the two Gal4-activating regions, also binds the inhibitor Gal80 (see Figure 2.4). These two functions can be distinguished by mutation: many single-amino-acid substitutions in ARII decrease or abolish the interaction with Gal80 while having little or no effect on the activation function. These results suggested that the same stretch of peptide can form two alternative structures, one rather articulated structure required for binding Gal80, and another less-defined structure required for activation.[11]

New Activating Regions Are Easily Generated

Many different peptides, when attached to a DNA-binding domain (e.g., that of Gal4 or LexA), work as activating regions. Most of these new acti-

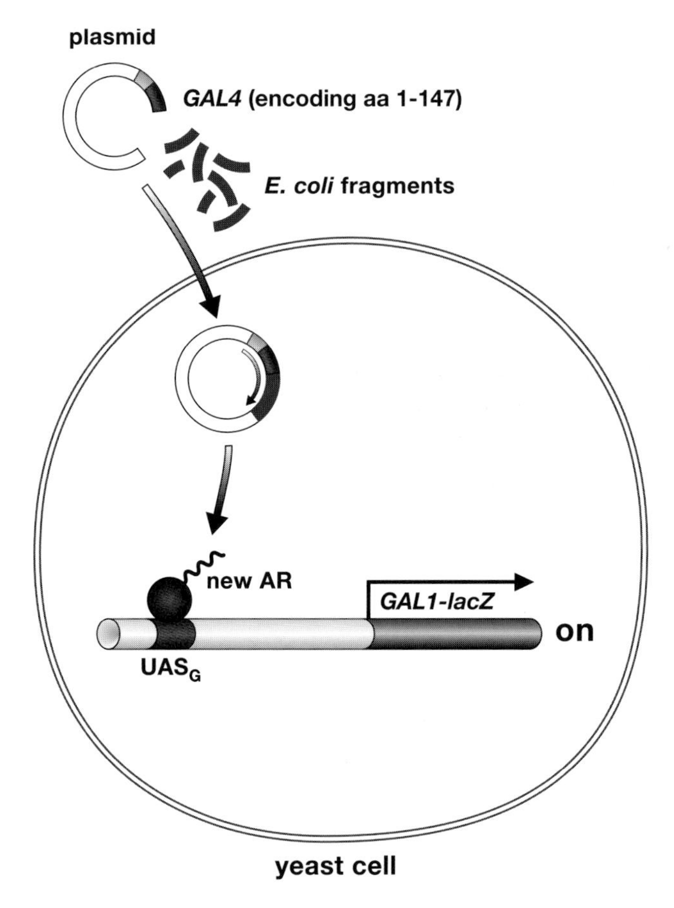

FIGURE 2.9. Isolation of new activating regions (ARs). Splicing random bits of *E. coli* DNA to a fragment encoding a DNA-binding domain (Gal4 [1–147]), produces genes that encode new activators. These genes are isolated at a high frequency (~1% of the random fragments tested work). Those that work are readily detected by a colony color assay; when the *GAL1-lacZ* reporter gene is activated, and β-galactosidase made, an indicator in the medium (X-Gal) turns the yeast colony blue. The new genes are expressed from the strong promoter of the *Adh* gene, indicated here in gray. (Modified, with permission, from Ptashne 1992 [see Chapter 1].)

vating sequences resemble natural activating regions in that they are acidic and include hydrophobic residues. One way to isolate new activating regions is shown in Figure 2.9.[12] Another way to generate a strong activating region is to create a peptide comprising reiterations of an octamer sequence found in a natural activating region.

For these new activating regions, as for natural ones, scrambling the order of amino acids destroys function. Thus, the presence of acidic and hydrophobic residues alone does not suffice—their arrangement in the sequence is also important. These, together with the other properties of activating regions we have discussed, suggest that they contain reiterated motifs, albeit motifs that are difficult to recognize.

Many of the properties of activating regions are shared by topogenic sequences, peptides that mark proteins for transport in and out of the nucleus. Topogenic sequences are believed to perform simple adhesive functions. (see Appendix 2, Topogenic Sequences).

Gal4 has properties that we would thus expect of an activator that works by recruitment. The following experiments support the idea that indeed this is how it works.

Squelching

Activators that work by recruitment cannot activate without binding to DNA. If not bound to DNA, the activator would be unable to direct the transcription machinery to a gene. Indeed, if it had any effect, it would be negative. For example, as shown in the Figure 2.10, at high concentration, an activating region (carried in this case on protein A) would titrate target sites on the transcriptional machinery and thereby make them unavailable for fruitful interaction with a DNA-bound activator (in this case B). Such

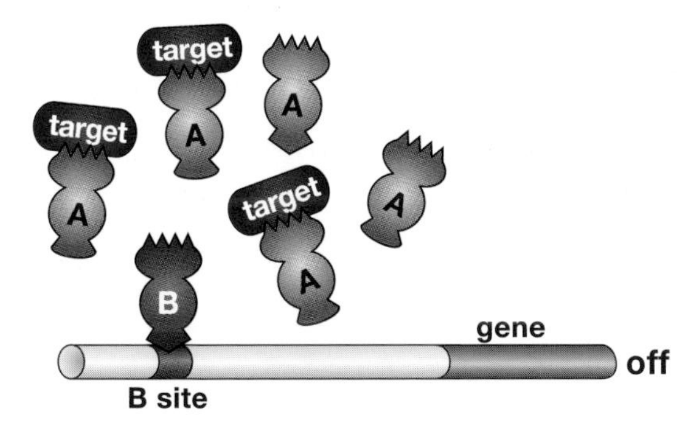

FIGURE 2.10. Squelching. An excess of activator A inhibits activator B from working. This is true for activators that work by recruitment. In this depiction, A has a DNA-binding domain, but the gene has no binding site for it. Squelching can also be observed by overexpressing an activating region lacking a DNA-binding domain. If B itself were expressed at a sufficiently high level (in the absence of A), it would "self-squelch." (Modified, with permission, from Ptashne 1992 [see Chapter 1].)

an effect would be more likely the stronger activating region—and thus the tighter it bound its target—and the more highly it was expressed. In contrast, and as we saw in our discussion of bacteria, activators that work like NtrC, when overexpressed, have the opposite effect: they promiscuously activate genes bearing a prebound polymerase.

Which class of activator does Gal4 resemble in this regard? Overexpression of a Gal4 fragment lacking a DNA-binding domain does not activate transcription. And, if the *GAL* genes are being activated by a DNA-bound activator, their expression is inhibited. Similar results have been observed with other yeast activators. In general, the stronger the activating region, and the more highly overexpressed the activator, the greater the degree of inhibition. The phenomenon is called *squelching*.[13]

It is this effect that we believe accounts for the generalization mentioned above that one rarely finds activators free of inhibitors in the nucleus, except when they are working appropriately. Were this not the case, there might be such a high concentration of activating regions in the nucleus that transcription in general would be squelched. Similarly, were repressors such as Mig1 to be overproduced in the nucleus, they might titrate the recruiting sites on the Tup1 complex and squelch (relieve) repression.

Recruitment Visualized

Again consider two alternatives that we encountered in *E. coli*: at genes activated by CAP, polymerase appears at the promoter, stably bound, only upon activation; in contrast, NtrC activates a polymerase prebound to a promoter.

The experimental technique called ChIP (chromatin immunoprecipitation) reveals the identities of proteins bound to specific regions of DNA in vivo. ChIP experiments show that the polymerase and associated proteins appear at the *GAL1* promoter only upon activation of the gene by Gal4 (see Figure 2.11). We return to this experiment below (see page 85, Action of Gal4: Experiments Performed In Vivo).

Activator Bypass Experiments

We saw that bacterial genes activated by recruitment can be activated in a variety of "activator bypass" experiments. In such experiments, the ordinary activator (e.g., CAP or λ repressor) is dispensed with, and activation is achieved by other means that increase the concentration of the poly-

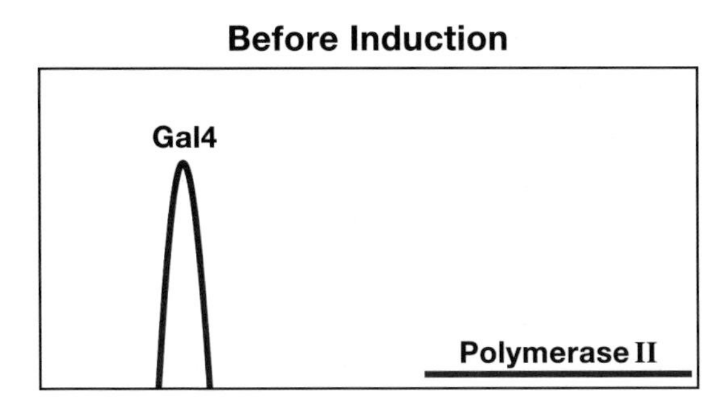

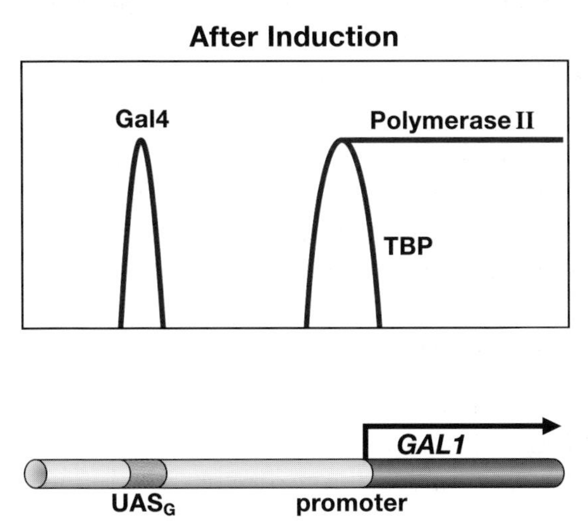

FIGURE 2.11. Recruitment visualized by ChIP analysis. (*Top panel*) Gal4 is bound at the UAS$_G$ before induction, but neither TBP nor PolII is found at the gene to any significant extent. (*Bottom panel*) Situation following addition of galactose (which frees Gal4's activating region as we have described). In this case, Gal4 is still found at the UAS$_G$, but now TBP is found at the promoter, and PolII is found both there and throughout the gene. Other components of the transcriptional machinery also bind upon induction but are not shown here. The technique used here has a resolution to about 200 bp, and so distinguishing between the UAS and the promoter (for example) is more difficult than implied by this schematic representation.

merase at the gene. Such an effect would not be expected for genes bearing prebound polymerases, e.g., those activated by NtrC and MerR. Here are two activator bypass experiments performed with the yeast *GAL1* gene, each analogous to an experiment that we described in bacteria.

Activation by a Heterologous Protein-Protein Interaction

A protein-protein interaction between a DNA-bound protein and the transcriptional machinery—one created by a mutation—can suffice for activation. This experiment is analogous to that of Figure 1.8, in which a heterologous protein-protein interaction substitutes for the ordinary interaction between CAP or λ repressor and *E. coli* RNA polymerase.

Recall that Gal4(1–100) dimerizes and binds DNA efficiently but, because it lacks an activating region, it does not activate. A mutant yeast strain was isolated in which this fragment activated transcription as efficiently as intact Gal4. The mutant strain bears a single-amino-acid change in a protein called Gal11 to form Gal11P (potentiator) (see Figure 2.12a,b).

Binding experiments performed in vitro show that the "P" mutation has created a protein-protein interaction between Gal11 and Gal4(1–100). Structural and other studies show that Gal11P interacts with a hydrophobic patch on the Gal4 fragment. Gal11 is a component of the mediator, one of the complexes of the transcriptional machinery that we alluded to at the beginning of this chapter (see Table 2.1).[14]

Two results support the notion that this new interaction, between Gal11P and the DNA-binding domain of Gal4, is simply adhesive.

- The interacting components can be swapped without loss of activity. Thus, a fragment of Gal11 bearing the P mutation, fused to the LexA DNA-binding domain, activates transcription, provided Gal4(1–100) (which includes the dimerization region) has been fused to Gal11 itself (see Figure 2.12c).
- The original Gal11P phenotype was conferred by substitution of an isoleucine for an asparagine residue at position 342 of Gal11. Other

FIGURE 2.12. Activator bypass I: The Gal11P effect. (*a*) Gal4(1–100) does not activate in wild-type cells; this DNA-binding fragment does not bind the mediator/holoenzyme. (*b*) The "P" mutation in *GAL11* creates an interaction with the Gal4 fragment and that interaction results in activation. Gal11 is a component of the mediator and so presumably that complex is directly recruited. Other components of the transcriptional machinery must also bind the promoter but are not shown in the figure. (*c*) The interacting components from parts *a* and *b* have been swapped. A fragment of Gal11 (comprising less than 10% of the molecule but including the P mutation) is fused to the LexA DNA-binding domain. Gal4(1–100) is fused to a different fragment of Gal11, a part that interacts with the mediator.

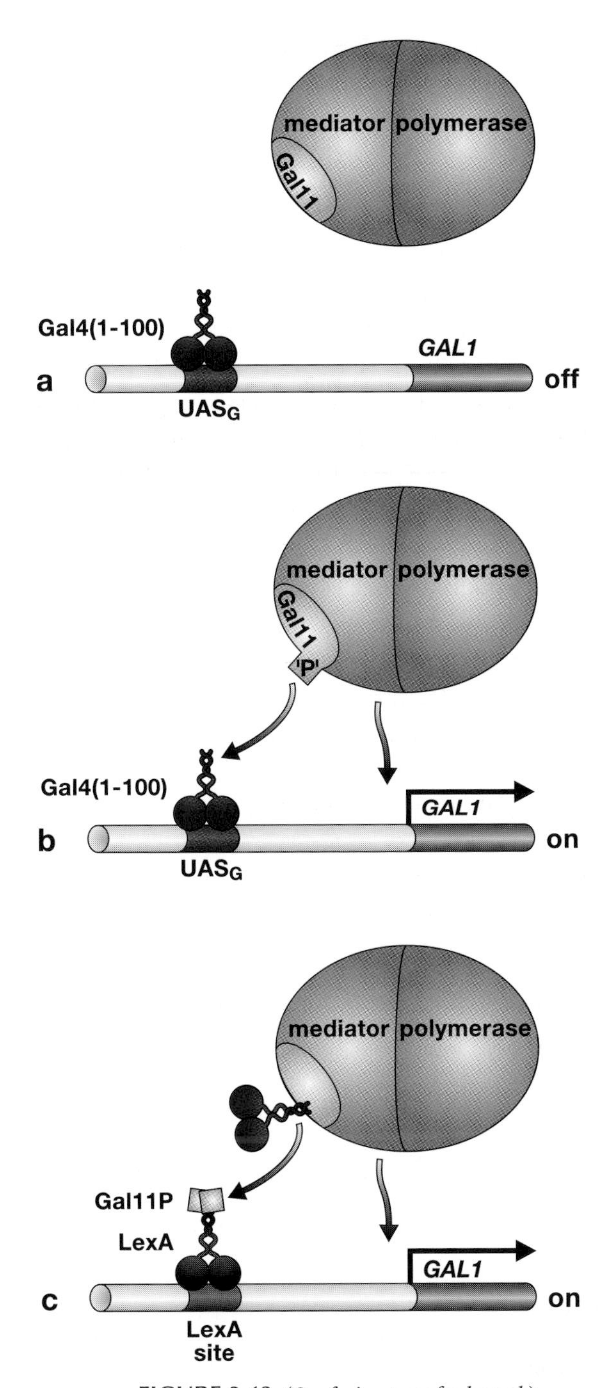

FIGURE 2.12. (*See facing page for legend.*)

hydrophobic residues introduced at this position also impart the Gal11P phenotype, but to a greater or lesser degree. The strength of the phenotype in turn predicts the relative strength of the interaction with Gal4(1–100) as measured in vitro.[15]

Direct Tethering of the Transcriptional Machinery

In Figure 1.9, we saw that directly fusing a subunit of *E. coli* RNA polymerase to a DNA-binding domain is sufficient to activate transcription of a gene bearing the corresponding DNA site nearby. The analogous result is observed in yeast: cells bearing the fusion protein LexA-Gal11 express the *GAL1* gene provided a LexA-binding site has been inserted nearby (see Figure 2.13). Likewise, when Gal11 is fused to the DNA-binding domain of Gal4, it activates transcription, provided Gal4 sites are present near the gene.

Activator bypass experiments of this sort have been performed using other components of the transcriptional machinery fused to DNA-binding domains. Some but not all such fusion proteins work, and the stimulatory effect can depend on the arrangement of binding sites and the identity of

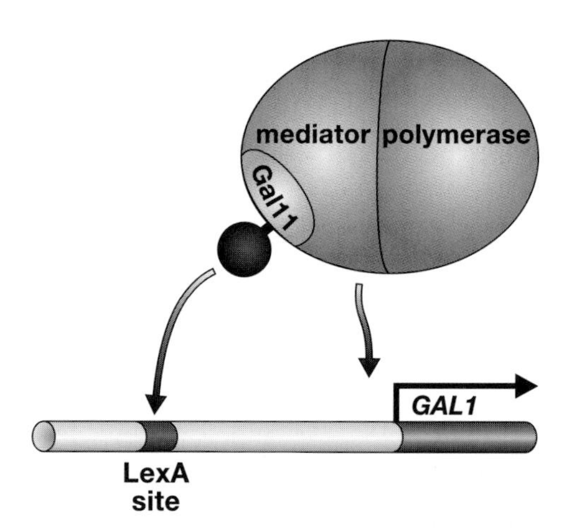

FIGURE 2.13. Activator bypass II: Direct tethering of the mediator/polymerase. LexA-Gal11 activates the *GAL1* gene in the absence of Gal4. The effect requires that LexA-binding sites be introduced near the gene. Only one LexA DNA-binding domain is shown; presumably there are two, because LexA monomers bind DNA only very weakly. Perhaps Gal11 is itself a dimer.

ACTIVATOR BYPASS EXPERIMENTS

Throughout the text we describe various so-called activator bypass experiments. Here, we summarize the nature, importance, and limitations of such experiments. Having established that a transcriptional activator has two functions, one ascribed to its DNA-binding domain and the other to its activating region, we are faced with the problem of what the activating region does. Finding that the activating region touches the polymerase, or some other components of the transcriptional machinery, only partially answers the question. The activating region might induce a critical conformational change in its target, or it might merely recruit the target to a nearby promoter, or, of course, it might do both.

As we have stressed, the resolution of this question has important implications for the nature of activating regions and how they interact with their targets. In one version of the activator bypass experiment, the ordinary polymerase-activating region interaction is replaced by some unrelated protein-protein interaction. One protein, tethered to DNA, interacts with another that has been attached to a component of the transcriptional machinery. As we have seen, such experimental manipulations can elicit high levels of expression of certain genes in bacteria and yeast. In another kind of activator bypass experiment, a DNA-binding domain is fused directly to the transcriptional machinery and, provided the appropriate DNA-binding site is introduced near a promoter, high levels of transcription can be achieved.

A positive result in such an experiment shows that simple tethering of the machinery (which we have referred to as recruitment) *suffices* for high-level expression of the gene. Taken alone, however, it does not show that the natural activator necessarily works that way. And there is always a problem of quantitation: one must determine whether, for any given case, the bypass construct is indeed working efficiently. Thus, the results of activator bypass experiments—like those of any other single kind of experiment—must be taken with other observations before a strong conclusion can be drawn. The many additional arguments that we have adduced—including the nature of activating regions, the requirement for DNA binding for activation, and the observation of recruitment of the machinery to promoters upon activation, etc.—are all consistent with the simplest inference of the bypass experiments, namely, that many activators work by recruitment.

More generally, wherever a natural interaction can be replaced by a different one, the result can be highly suggestive of function. For example, the fact that the Gal4 DNA-binding domain can be replaced by a bacterial DNA-binding domain, without loss of function (provided the appropriate DNA-binding site is provided; see page 69) suggests that the sole role of DNA binding is to suitably position the activating region near the gene to be activated. And the fact that a natural repressor (Mig1, see below) can be replaced by a Lex-Tup1 fusion suggests that the sole role of Mig1 is to recruit the repressing complex.

the promoter tested. Gal11 fusions generally work more efficiently and reliably than the others. A possible explanation would be that Gal11 is a component of the mediator and can interact with other components of the machinery as well.[16]

The activator bypass experiments show that bringing the machinery to the *GAL1* gene can suffice for activation. This result—taken with the nature of activating regions, the squelching phenomenon, and the results of ChIP analysis—argues that Gal4 works by recruitment. And in particular, these findings strongly argue against the two alternative mechanisms for activation that we encountered in bacteria.[17]

How Does Gal4 Recruit Polymerase?

We are now faced with the question of how recruitment is effected. In bacteria, the answer was straightforward: polymerase is the only relevant complex for the activator to recruit. In yeast, as we noted, we have several additional complexities: the machinery that transcribes genes comes in parts; the DNA is wrapped around histones to form nucleosomes; and various complexes chemically modify nucleosomes or alter their disposition on DNA. We start this discussion by considering the effects of nucleosomes and their modifiers.

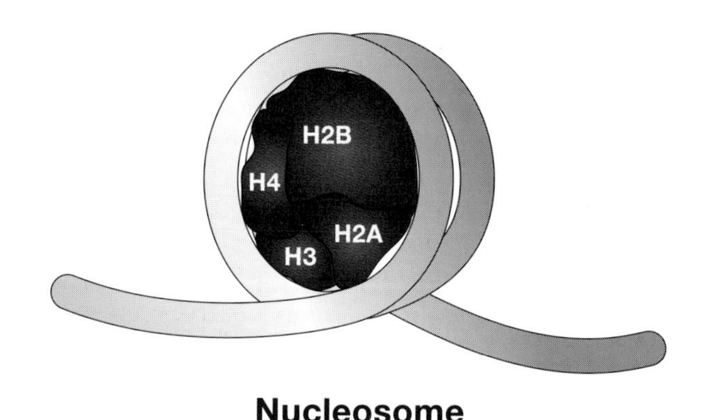

FIGURE 2.14. A nucleosome. A dimer of the four depicted histones comprises the histone octamer. DNA wraps around the octamer in slightly less than two turns. The linker DNA between adjacent nucleosomes varies from 15 to 55 bp in different organisms. In addition to the amino-terminal tails mentioned in the text, the histone H2A also has a tail at its carboxyl terminus.

Nucleosomes and Their Modifiers

Figure 2.14 shows a schematic representation of a nucleosome. Each nucleosome (as in all eukaryotes) bears 146 bp of DNA wrapped around the histones. Each histone has an extended 15–25-amino-acid tail at its amino terminus. The tails are not shown in the figure, and their disposition is not known. Each nucleosome is separated by 15–20 bp, called linker DNA. Two further observations are particularly relevant here.

- As assayed in vitro, nucleosomal DNA (but not linker DNA) is less accessible than naked DNA. For example, certain nucleases (enzymes that cut DNA) work orders of magnitude less efficiently on nucleosomes than on naked DNA.
- Actively transcribed genes are often associated with modified nucleosomes: the tails of histones H3 and H4 are typically acetylated at specific lysine residues, for example. HATs (enzymes, recall, that acetylate histones) have been shown, by genetic analysis in yeast, to be required for full induction of certain genes; HDACs (enzymes that deacetylate histones) have been implicated in repressing certain genes.

Two plausible interpretations of the preceding facts are as follows:

- Nucleosomes present a barrier to the transcriptional machinery (and to DNA-binding proteins in general), and whereas acetylation of the tails of certain histones weakens that barrier, deactylation strengthens it. One version of this scenario would be that acetylation of lysines in the histone tails eliminates positive charge and thereby decreases interaction of those tails with the negatively charged DNA.[18]
- Acetylation directly facilitates protein binding. Certain proteins found in the transcription machinery bear so-called bromo domains that bind to acetylated lysines in histone tails. Examples are the HAT Gcn5 and one of the proteins associated with TBP (TAF145). It is possible that modified nucleosomes have a higher affinity for the transcriptional machinery than even naked DNA does. Other modifications (e.g., phosphorylation) also create docking sites for other proteins.[19]

Gene expression can also be affected by enzymes (e.g., the so-called Swi/Snf complex) that alter the mobility or configuration of histones on DNA. The resulting nucleosome fluidity can help certain proteins to bind DNA.

The various enzymes that affect nucleosome structure or disposition work without any DNA sequence specificity. The background activities of HATs and HDACs produce a steady-state level of about 12 acetyl groups per nucleosome, spread over the yeast genome. Some of these histone-modifying enzymes may act on other substrates as well, including regulators and parts of the transcriptional machinery.[20]

In light of these observations, consider the following scenarios for how activators might work. At one extreme, direct recruitment of the transcribing complex would suffice for the initiation of transcription. At the other extreme, recruitment of nucleosome modifiers would suffice, with the transcribing complex then forming without additional help. Yet another alternative would be that both kinds of complexes are directly recruited.

The following studies of activation by Gal4 and related activators lead to the following conclusion: the typical acidic activator directly recruits those proteins that are involved in transcribing the gene and, where required, nucleosome modifiers as well.

Targets of Gal4: Experiments Performed In Vitro

Gal4's activating regions, and other acidic activating regions as well, interact with several components of the transcriptional machinery when tested by binding assays peformed in vitro. These include components of the mediator (including Srb4 and Srb10), TBP, TFIIB, TFIIH, TAFs, as well as various nucleosome-modifying complexes including Swi/Snf and SAGA.

The following experiments show that at least some of these interactions can be functionally significant in vitro. These experiments demonstrate recruitment of various complexes to DNA templates, naked or wrapped in nucleosomes.

- On a DNA template lacking nucleosomes (i.e., naked DNA), transcription was activated by Gal4 (or a similar activator) in a reaction unaffected by nucleosome modifiers. Activation required the mediator and other factors required generally for transcription including the polymerase (see Table 2.1). We do not know which components are directly recruited and which bind cooperatively along with those.

  Following the initiation of transcription, some of those components, including TBP and the mediator, remained at the promoter. It is suggested that this "scaffold" can be stabilized by the activator, and subsequent rounds of initiation require recruitment (or spontaneous binding) of a more limited set of factors than was required initially.

- Using DNA templates assembled into nucleosomes, and assaying directly for nucleosome modifications, nucleosome modifiers—both Swi/Snf and SAGA—were found to be recruited.
- In a transcription experiment using nucleosomal templates, conditions were found under which transcription required direct recruitment of both nucleosome modifiers and the transcribing machinery.[21]

For further comments on interpreting in vitro experiments, see Footnote 22.

Action of Gal4: Experiments Performed In Vivo

The experiments described above show that acidic activators can recruit different kinds of complexes to templates in vitro. We now turn to experiments involving activation of *GAL1* by Gal4 in vivo and ask two questions: What components of the transcriptional machinery are required for activation? Which components actually appear at the *GAL1* gene upon activation? We begin by briefly outlining three experimental strategies that help answer these questions. Footnotes comment further on each approach.

- ***ChIP (chromatin immunoprecipitation):*** This method, as we have mentioned, can detect proteins bound to defined regions of DNA in cells. It can also detect enzymatic modifications—acetylation of nucleosomes, for example.[23]
- ***Genetic deletion:*** Genes encoding components of the transcriptional machinery can, in many instances, be deleted without killing the cell. One can ask whether, in a strain lacking one or another of these components, expression of a given gene is affected.[24]
- ***Protein destruction:*** Components that cannot be deleted (by mutation) without loss of viability can nevertheless be eliminated from cells by a variety of methods, and the effect on expression of a given gene can be studied in the dying cells.[25]

These experimental strategies have been applied to activation of the *GAL1* gene by Gal4 and have provided the following conclusions.

What Is Required for Activation?

- Deletion of either the HAT Gcn5 or of the nucleosome modifier Swi/Snf has little if any effect on activation.
- Deletion of some mediator components—Gal11, Srb10/11—modestly decreases activation (about two- to fivefold), and removal (by protein destruction) of most TAFs has a small if any effect.
- Deletion of certain SAGA components (other than Gcn5) decreases activation more dramatically.
- The polymerase, the general transcription factors (GTFs), and some mediator components (e.g., Srb4) are absolutely required, as they are at almost all genes.

What Appears at the Gene Upon Activation?

As mentioned earlier, induction of the *GAL1* gene in wild-type cells results in the appearance of various components of the transcriptional machinery at the promoter (see page 76 above, Recruitment Visualized). These components include the polymerase, TFIIH, TFIIE, TBP, various mediator components, and SAGA (including the TAFs that are part of SAGA).[26]

Certain components are found only at the promoter (e.g., TFIIE and mediator), whereas others are found both at the promoter and downstream in the gene itself along with polymerase. The result is consonant with an experiment performed in vitro and described above: part of the machinery remains at the promoter following initiation.

In sum, components of the mediator and SAGA contribute to activation of *GAL1* by Gal4. Both complexes appear at the promoter upon activation, along with polymerase and, to the extent assayed, GTFs. All of these complexes (except polymerase, evidently) might be recruited directly, or some might bind cooperatively along with those recruited directly. Reinitiation might involve recruitment of a subset of those factors recruited for the first round of transcription.

Why is SAGA required in view of the finding that its acetylating activity (Gcn5) is dispensable? The answer may be that the complex (even without Gcn5) helps recruit TBP: one subunit of SAGA interacts with TBP and that interaction is required both for activation and for stable recruitment of TBP.

The following experiment is consistent with the idea that in activating efficiently, Gal4 touches both SAGA and the mediator. A fusion protein comprising a component of SAGA attached to a DNA-binding domain elicits very little, if any, activation on its own. Similarly, as we pointed out earlier, some fusion proteins bearing mediator components also activate very weakly. But when binding sites for both of these fusions are placed in front of *GAL1*, strong activation is observed.

Imposing a Need for Nucleosome Modifiers

We have seen that neither the HAT Gcn5 nor the remodeler Swi/Snf is strongly required for activation of *GAL1* by Gal4. But the expression of some genes is far more dependent on these activities. In these cases, which constitute about 5% of genes, deletion of either significantly reduces expression.[27]

What is the difference then, between those cases where the modifiers are required and those where they are not? The answer is suggested by the finding that a requirement for these modifiers can easily be imposed on the action of Gal4 at the *GAL1* gene, as we now describe.

Weakening the Gal4-binding Sites or the Gal4-activating Region

If the four Gal4-binding sites in the UAS_G are replaced with two weaker binding sites, mutation of either Gcn5 or Swi/Snf decreases transcription of *GAL1* at least tenfold (see Figure 2.15a). Gal4 working from two strong sites does not require the modifiers. In addition, a derivative of Gal4 bearing a weakened activating region now depends on Gcn5 when working from a UAS_G.[28]

These results suggest that the need for modifiers depends on an energetic factor. Apparently, if Gal4 binds its sites sufficiently tightly, interacts with components of the transcribing machinery sufficiently strongly, and that machinery binds the *GAL1* promoter sufficiently avidly, additional help from nucleosomal modifiers is unnecessary. Weakening one or another of the interactions can impose a requirement for these modifiers.

There is a state of the cell in which induction of the *GAL* genes, even in the wild-type configuration, depends on both Gcn5 and Swi/Snf as we now describe below.

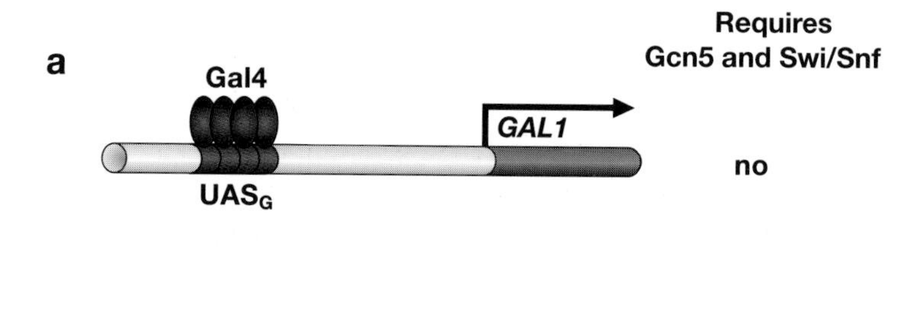

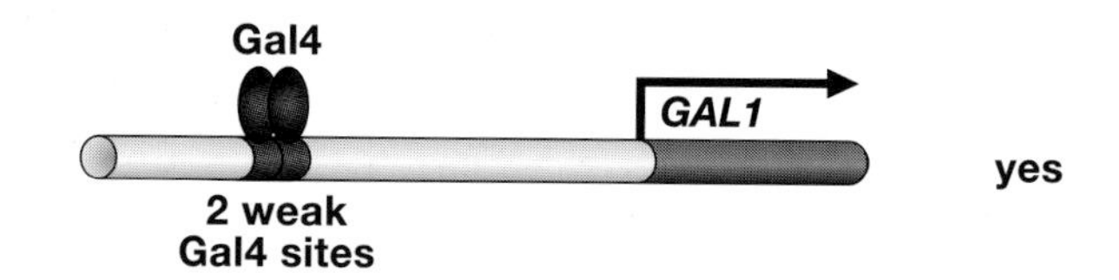

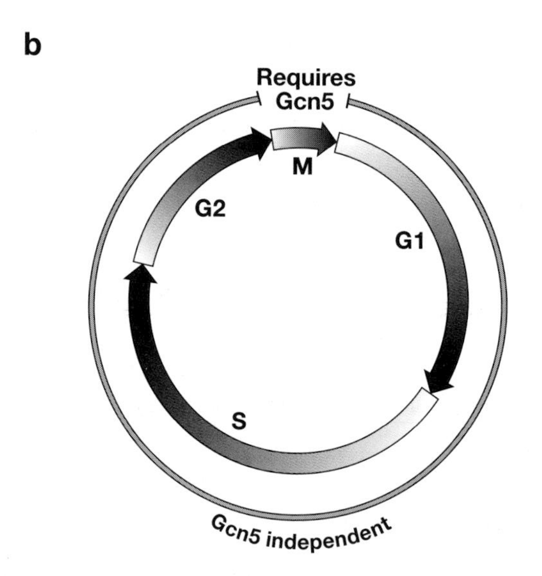

FIGURE 2.15. Two ways to impose a requirement for nucleosome modifiers on the *GAL* genes. (*a*) UAS$_G$ in the first construct has been replaced, in the second construct, by two weak Gal4-binding sites (sites 1 and 2 from the UAS$_G$). The cells carrying these reporters are assayed as they grow asynchronously. (*b*) Cells are synchronized, and induction of a gene bearing a UAS$_G$ (such as that in the top line of part *a*) requires Gcn5 (and Swi/Snf) during the M (mitosis) phase of the cell cycle. S stands for DNA synthesis, and G1 and G2 are characteristic gaps between these phases of the cell cycle.

Cell-cycle Stage-dependent Requirement for Nucleosome Modifiers

The state of the DNA differs at different stages of the cell cycle. In particular, chromosomes are condensed during mitosis, a relatively short part of the total cell cycle. The experiments involving living cells that we have discussed thus far all used asynchronously growing cells, and thus any stage-specific effect would probably have been overlooked.

Yeast cells can be synchronized to enter each stage of the cell cycle (e.g., mitosis) in concert, allowing induction of *GAL1* to be measured at any given stage. Such an experiment reveals that induction of *GAL1* during mitosis will proceed normally only if both Swi/Snf and Gcn5 are present. At this stage of the cell cycle, evidently, the chromatin (i.e., the DNA plus associated proteins) is in such a configuration that those enzymes are required. Once again, we do know whether these modifiers need to be recruited or whether they work as part of the background (see Figure 2.15b).

We thus surmise that there are two classes of genes whose expression ordinarily requires nucleosome modifiers.

- Certain genes transcribed during mitosis encode products that are required for efficient progression to the next stage of the cell cycle. Many of these genes require the modifiers for efficient expression.[29]

- Other genes expressed during different stages of the cell cycle might require the modifiers because they resemble, in one way or another, the weakened *GAL* system described in Figure 2.15a. Activation of the gene *HIS3* (which encodes a protein required for histidine biosynthesis) by its activator Gcn4 is an example of a case in which the nucleosome modifiers are sequenced.

Activation: An Interim Summary

Of the three modes of activation that we encountered in bacteria, two clearly do not apply to activation of the *GAL1* gene by Gal4. That is, Gal4 does not instruct prebound but inactive transcriptional machinery to work (in the way that NtrC does); and neither does it alter the DNA state to make it transcribable (in the way that MerR does). To the contrary, the characteristics of the Gal4 DNA-binding and activating regions; the appearance of the transcriptional machinery at the *GAL1* promoter upon activation; the phenomenon of squelching; and the success of activator bypass experiments are all consistent with the idea that Gal4 works by recruitment. We draw these conclusions despite our ignorance of many details, including the roles of most of the proteins found in the transcriptional machinery.

We do not mean to rule out the possibility that activators such as Gal4 might, in addition to recruiting the machinery, have other effects that help activation. For example, it has been suggested that TBP is engulfed by a TAF (TAF145), and interaction of an activating region with the underside of TBP (that which binds DNA) displaces from that surface a region of the TAF that acts as an inhibitory flap. How replacing an inhibitory flap with an activating region might help activation is not clear. Moreover, such an effect cannot be crucial to activation as evidenced by the success of activator bypass experiments. Nevertheless, such effects might occur and might help the activator to work more efficiently.[30]

In contrast to the cases of regulated recruitment in bacteria, it is difficult to be precise about what Gal4 directly recruits (touches) in any given case, or what it *must* recruit. We summarized above a likely scenario for activation of *GAL1* during most of the cell cycle: Gal4 touches both SAGA and the mediator. We noted that the requirement for nucleosome modifiers (in addition to SAGA), which obtains only under more restricted conditions, might be met by direct recruitment of those modifiers as well.

Gal4 could, by virtue of its multiple interactions, not only recruit the initial complex that binds at the promoter—and stabilize it there—but also re-recruit components that leave the promoter and move downstream with the transcribing polymerase.

Although we have not discussed this matter explicitly, recruitment can also stimulate transcript elongation. Recall that TFIIH, which has kinase and helicase activities, is essential for the escape of polymerase from the promoter. TFIIH, as we have also noted, binds acidic activating regions in vitro, and hence may also be recruited by Gal4. In the following chapters, we will encounter a kinase that is required for transcript elongation, and we will see that that kinase is recruited by certain activators.

The apparent ability of any given activator (e.g., Gal4) to directly recruit multiple complexes (nucleosome modifiers plus mediator, etc.) raises additional mechanistic problems. Can an activator directly recruit multiple targets simultaneously, or must recruitment proceed stepwise? If the latter, then does the activator necessarily dissociate from the DNA and rebind for each step? We return to this question in Chapter 3.[31]

Gene activation is a quantitative matter. The degree of activation elicited by an activator depends on many factors, including the strength of the activating region, the number of activator binding sites, and so on. Manipulations that weaken the recruiting reaction can impose a requirement for nucleosome modifiers.[32]

Repression by Mig1

Recall that in bacteria, repression does not involve recruitment. When working as a repressor, for example, λ repressor binds DNA and competitively inhibits binding of polymerase. Many eukaryotic regulators work differently: they bind DNA and recruit a "repressing complex." Mig1 is an example.

Mig1, as we have noted, binds to DNA sites near the *GAL1* promoter and mediates glucose repression by recruiting the Tup1 repressing complex (see Figure 2.2c). That complex, which also includes the protein Ssn6, works at many different genes and is recruited by a variety of DNA-binding proteins.

The sole role of Mig1 is to recruit the repressing complex, as shown by the following Mig1 bypass experiments. A LexA-Ssn6 or a LexA-Tup1 fusion protein represses transcription when bound to suitably positioned LexA sites. The LexA-Tup1 fusion represses in a strain deleted for Ssn6, but the LexA-Ssn6 fusion requires Tup1, a result indicating that Tup1 bears the repressing function (Figure 2.16).

Two mechanisms have been suggested to explain Tup1-mediated repression.

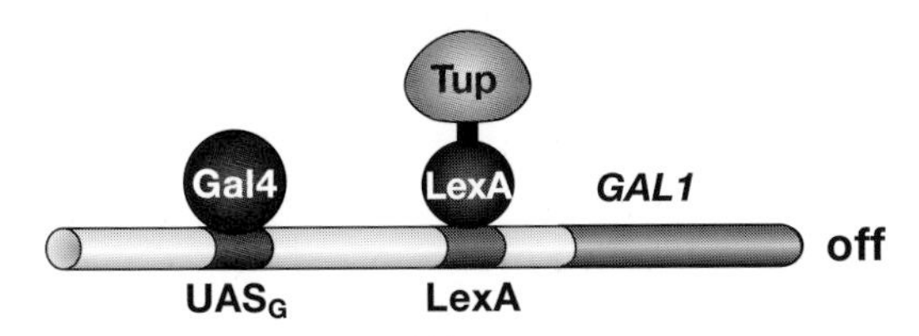

FIGURE 2.16. Recruitment of a repressing complex. LexA fused to the Tup1 site works as a repressor. Increasing the number of Mig1-binding sites in the reporter of Figure 2.2, or LexA sites in this reporter, increases repression. Tup1 also represses when tethered (or recruited) upstream of the activator, but less well.

- Tup1 recruits complexes (HDACs) that remove acetyl groups from histones, thereby rendering the promoter less accessible to the transcribing machinery. Consistent with this idea, inactivating three HDACs by mutation (but not of any one separately) severely compromises Tup1-mediated repression. This result suggests that Tup1 can recruit any of several HDACs to effect repression.
- Tup1 interacts with, and poisons, the transcriptional machinery at the promoter. Part of the evidence for this view is that deletion of certain components of the mediator eliminates the effect of Tup1 at certain genes.

Perhaps both of these mechanisms are used. If so, we would have a parallel between the action of activators and repressors: each would interact both with the transcribing machinery (mediator, GTFs, etc.) and with nucleosome modifiers. An important difference, however, would be that whereas all interactions of the activator serve recruitment functions, the interaction between Tup1 and the transcriptional machinery (if significant) would have to inhibit functioning; presumably, Tup1 interacts in a special way with its target to effect that repression.

In sum, repression, like activation, also involves recruitment: In this case, the Tup1 "repressing complex" is recruited to DNA by the specific DNA-binding protein Mig1. How the Tup1 complex works, once recruited, is not yet clear. There is evidence that it directly interacts with and poisons the transcriptional machinery and that it recruits HDACs.

Other repressing complexes include HDACs—Sin3-Rpd3, for example—and are recruited to genes by DNA-binding proteins other than Mig1. We will encounter HDACs that are involved in a form of repression called "silencing" below.

SIGNAL INTEGRATION AND COMBINATORIAL CONTROL

Signal integration ensures that certain genes are switched on (or off) only when the proper combination of two or more signals is present. As we have seen, the combination of an activator and a repressor can be used to integrate two signals in bacteria and in yeast, as at the *GAL* genes in both

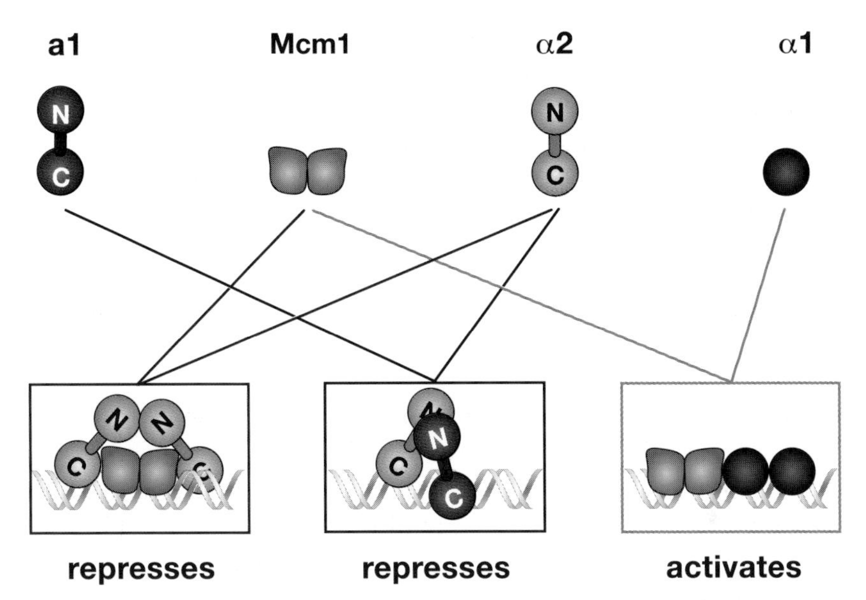

FIGURE 2.17. Combinatorial control: Alternative binding partners. Mcm1 works both as an activator and as a repressor. Mcm1 bears an activating region that is exposed when this protein binds with α1 but is covered when it is bound (at a different site) with α2. α2, when bound either with Mcm1 or with **a**1, recruits the Tup1 repressing complex. (Modified, with permission, from Ptashne 1992 [see Chapter 1].)

organisms. We have also seen that the same activator can work in combination with different repressors, e.g., CAP with the Lac and Gal repressors in *E. coli*, an example of combinatorial control.

We describe here two further examples of signal integration and combinatorial control in yeast. In each case, binding to DNA of one protein depends on another protein. But quite different mechanisms are used in the two cases.

Cooperative Binding with Alternative Partners

As shown in Figure 2.17, four proteins bind to DNA in different pairs. In each case, the binding is cooperative, and no one of the proteins, when present at physiological concentrations, binds to any of these sites on its own.

Two of these combinations repress, and the other activates. The same protein—Mcm1—activates in one case and represses in another. A complete description of the biological context in which these pairs bind is beyond our scope; a brief summary is found in Footnote 33.

It may seem surprising that a given regulator can interact (and bind cooperatively) with alternative partners. But as only weak interactions between binding partners are required for cooperativity, such interactions make only modest demands on protein structure. In only a few cases have interactions that mediate cooperative binding of the sort described here been visualized by crystallography. As anticipated, the protein surfaces involved are not highly complementary.

Sequential Binding of Activators

In this example, the binding of one regulator to DNA facilitates binding of another downstream, but in this case, the regulators do not touch each other. Rather, it is believed, the first recruits histone modifiers that allow the second activator to bind.

Expression of the gene called *HO* requires the two activators, Swi5 and SBF. There are multiple sites for each of these activators upstream of the *HO* gene; those for Swi5 are further upstream (see Figure 2.18). DNA binding of SBF depends on the prior DNA binding of Swi5. Swi5 recruits the Swi/Snf and SAGA complexes. These modifiers facilitate binding of SBF to its sites downstream. SBF, in turn, activates the gene.[34]

An obvious question arises: does Swi5 have some special property that enables it to recruit the nucleosome modifiers but not activate the gene? The answer is no: if the Swi5-binding sites are replaced by Gal4 sites, activation of the *HO* gene becomes dependent on Gal4 and SBF instead of on Swi5 and SBF.

Presumably Gal4, like Swi5, is not able to activate *HO* by itself because (in this case) it is bound too far away to recruit the transcribing machinery to the promoter. Gal4 can nevertheless recruit the nucleosome modifiers and thereby allow SBF to bind at its position, closer to the promoter, and activate the gene.[35] A summary of the biological context in which these events occur is given in Footnote 36.

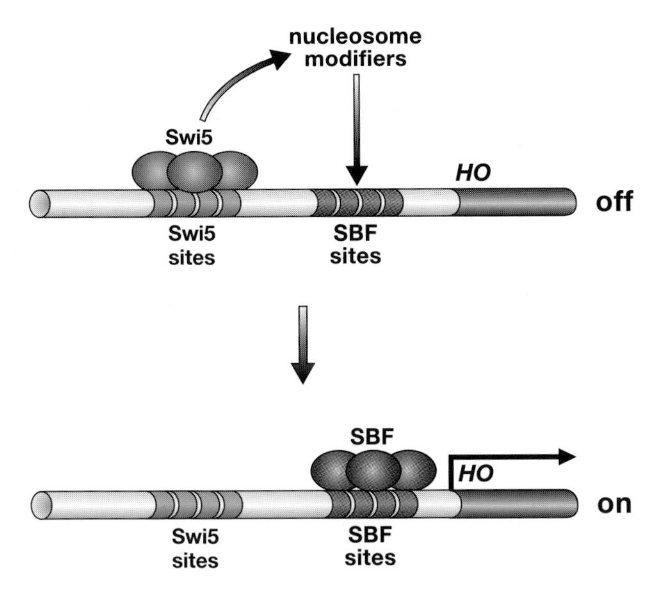

FIGURE 2.18. Control of the *HO* gene. It is not known how nucleosome modification extends from the Swi5 sites through the SBF sites. The modifiers might move along the DNA (some two nucleosomes' worth), or, because they are so large, even when tethered at Swi5 sites, their effects might extend to the SBF sites.

SILENCING

We encountered two broad categories of repression in bacteria. In the more common case, repressors bind to specific sites on DNA and repress transcription of specific genes. In the less common case, proteins that are bound to DNA *nonspecifically* coat a region of the chromosome, and this barrier decreases accessibility to other DNA-binding proteins. This latter phenomenon is called silencing (see More on Repression, Chapter 1).

Silencing also occurs in yeast. At each silencing region, one or more specific DNA-binding protein recruits other proteins that form the barrier, as we now explain.

Heterochromatic Chromosomal Regions

We will concentrate on silencing at DNA adjacent to the tip of a yeast chromosome, the telomere. There are three other silencing regions, all of which,

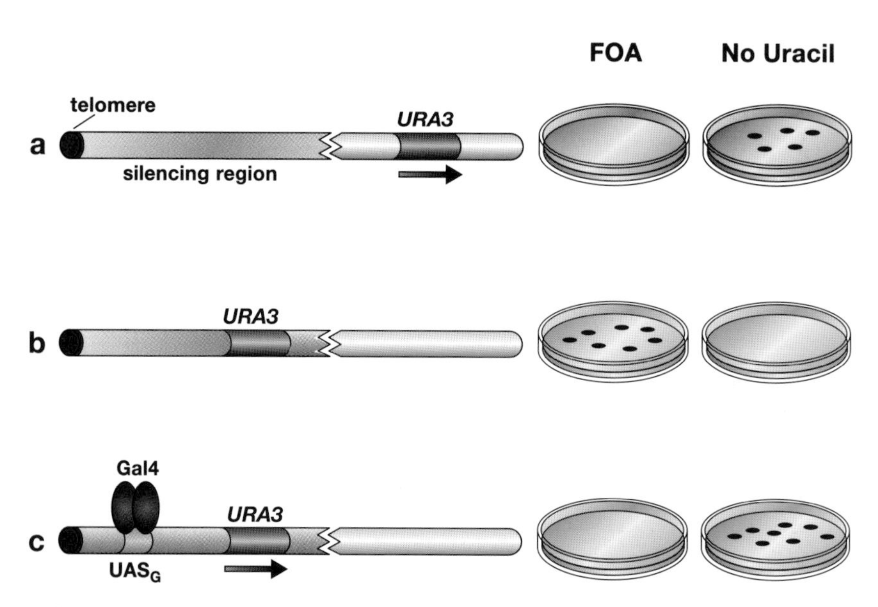

FIGURE 2.19. Silencing of a gene placed near a telomere. (*a*) When at its ordinary location away from a telomere, the basal level expression of the *URA3* gene (i.e., that observed in the absence of an activator) is sufficient to enable cells to grow on medium lacking uracil. That level of *URA3* converts 5-FOA in the medium to a poison that prevents cell growth. (*b*) If the gene (absent an activator) is moved near a telomere, it is silenced, and the cells grow on 5-FOA but not in the absence of uracil. (*c*) A strong activator (in this case Gal4) overcomes silencing.

like telomeric silencing, extend over some 2–3.5 kb. These three other regions are: the two "silent mating-type" (HM) loci and the locus that encodes multiple copies of the genes encoding ribosomal RNA.[37]

Silencing of a gene is most readily measured in the absence of an activator. For example, the *URA3* gene is activated in response to uracil depletion by the activator Ppr1. But even in the absence of Ppr1, the *URA3* gene, at its ordinary chromosomal location, produces enough product to enable cells to grow in the absence of uracil. When the gene (minus the activator) is moved to the telomere, however, it is silenced in many of the cells: these cells grow only if uracil is provided in the medium (see Figure 2.19).[38]

A strong activator bound near the gene overcomes the silencing. For example, if the telomere-linked *URA3* gene bears binding sites for Gal4, that activator elicits a high level of expression of the gene. Thus, there is a barrier to gene expression near the teleomere, but a barrier that is readily overcome. What then is the nature of that barrier?

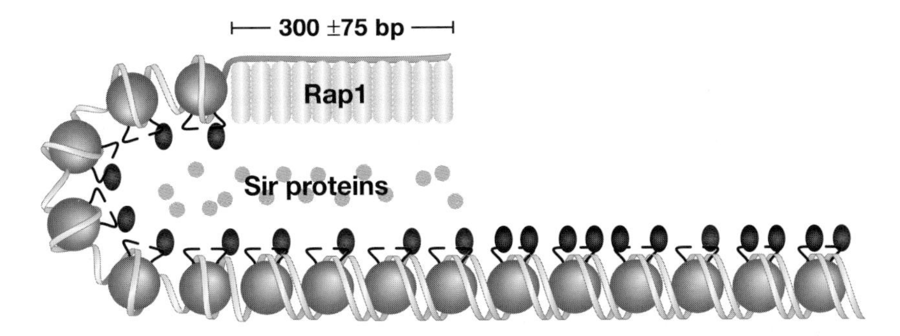

FIGURE 2.20. Folded structure of the end of the chromosome. Approximately 50 binding sites for the protein Rap1 are found at the telomere. This protein, in turn, recruits several so-called Sir proteins, including Sir2 (an HDAC), Sir3, and Sir4. The tethered Sir proteins evidently bind to nucleosomes to form the folded structure shown. Overexpression of certain Sir proteins can increase the size of the looped structure. Deacetylated lysines in histone tails are indicated in red, and acetylated lysines are indicated in black. (Modified, with permission, from Grunstein 1998 [copyright Elsevier Science].)

Each yeast telomere encompasses about 350 bp and includes binding sites for the protein Rap1 (see Figure 2.20). This protein, when bound, recruits a complex of several so-called Sir (silent information regulator) proteins. The Sir complex is believed to interact with subtelomeric regions, as shown in Figure 2.20, to form a folded or looped structure. Consistent with this picture, DNA near the telomere is more resistant to nucleases than is ordinary chromosomal DNA. These and other observations indicate that the DNA-protein complex near the telomere resembles "heterochromatic" chromosomal regions in higher organisms, regions also often associated with gene silencing.[39]

The silencing loci have a common requirement for the protein Sir2. Sir2 is an HDAC (histone deacetylase), the kind of enzyme we have discussed in the context of other examples of repression. Recruitment of Sir2 is achieved in a variety of ways. For example, at the rDNA locus, Sir2 is recruited by interaction with a DNA-binding protein (called Net1), whereas a different protein recruits Sir2 to the HM loci.[40]

Not only are there various ways to recruit Sir2 to the various silencing regions, but Sir2 can itself be replaced by another recruited protein. In cells lacking Sir2, silencing does not occur. But a single point mutation in these deficient cells has been shown to restore silencing at the HM loci. The new mutation evidently creates a strong interaction between a protein bound near the HM loci and another protein called Sum-1. The latter then binds and recruits Hst1, a relative of Sir2, which presumably performs the HDAC function, and silencing is restored.

The study of silencing has been revealing in two additional ways, matters that we now briefly describe.

Compartmentalization

We have seen how moving a gene from one position on the chromosome to another can affect its expression. Here we encounter an effect of changing the physical location of the gene in the cell.

At the HM loci, three DNA-binding proteins (the identities of which need not concern us) bind to three separate DNA sites and recruit the Sir proteins. In the absence of two of these sites, silencing is greatly reduced. Repression is restored to such a weakened HM locus if Sir2 is artificially overproduced in the cells.

Another way to restore silencing to the weakened HM locus is to move it to the nuclear membrane. That was achieved using a hybrid protein designed to move DNA to the membrane. The hybrid protein comprises a Gal4 DNA-binding domain fused to a protein that imbeds in the membrane. This hybrid protein restored silencing to a gene at the weakened HM locus provided a Gal4 site was introduced near that locus. Re-establishing silencing at the weakened HM locus—either by overexpressing Sir2 or by moving the locus to the membrane—requires the single remaining binding site for a Sir recruiting protein. In the absence of that site, neither manipulation will cause silencing.

The membrane tethering experiment suggests that Sir proteins are at an especially high concentration in the compartment defined as the region near the nuclear membrane. Consistent with that conclusion, both telomeres and Sir proteins have been judged from cytological studies to be clustered near the nuclear periphery.

It is believed that Sir proteins are concentrated at the nuclear periphery not because they spontaneously accumulate there, but rather because they are associated with telomeres that themselves are somehow anchored there. Evidently the silenced regions interact; that interaction is fostered by their physical proximity and helps stabilize the silencing structures. Consistent with this view, a damaged HM locus is silenced more efficiently when it is moved from its ordinary chromosomal region to a chromosomal position closer to the telomere.[41]

Variegation

There is more to silencing than silencing. We have emphasized that a gene, expressed weakly at its ordinary chromosomal location, can be effectively turned off when positioned near the telomere. But about once every 10–20 cell divisions, the silenced gene spontaneously switches on (to a low level). Then, about 10–20 generations later, it switches off again. We say, therefore, that expression of the gene is "variegated." Variegation in yeast is easily observed with the *ADE2* gene, as discussed in the panel.

> **VARIEGATION VISUALIZED**
>
> A yeast colony arises from a single cell; many generations produce the million or so cells found in a typical large colony growing on an agar plate. Should the cells contain a gene whose product confers a characteristic color, and if the color is restricted to the cell in which that protein is produced, then variegation is easily observed. Thus, if such a gene, lacking a strong activator, is placed near a telomere, the colony will be mixed: some cells will be colored, some uncolored, as gene expression variegates.
>
> Consider an example in which the cells bear an *ADE2* gene near the telomere. When the gene is on (even to a very low level) the cells are white, whereas when it is totally off, the cells are bright red. The red color arises from accumulation of the substrate of the enzyme encoded by *ADE2*. The typical colony bearing *ADE2* at the telomere is sectored: mostly red with white streaks. When cells are picked from the white sectors and replated, they give rise to new colonies, which are mostly white with red sectors.

We do not have a detailed understanding of the molecular events underlying variegation. A reasonable scenario might go as follows. The gene is initially silenced by the barrier presented by the silencing heterochromatin, as we have indicated. But that barrier is weak and it might be disrupted (occasionally) by the binding of the transcriptional machinery; that binding might occur spontaneously, or it might be helped by a weak activator.

So why, once it is switched on, does the gene tend to stay on for several generations? A reasonable explanation would be that once disrupted, reformation of the silencing heterochromatin, which presumably requires the binding of multiple proteins, is a relatively rare event. Moreover, the active state might be self-perpetuating: HATs associated with the transcriptional machinery could modify nucleosomes, which in turn would have an increased affinity for that machinery. Similarly, once the silenced

EPIGENETICS

The classical definition of epigenetics is "a change in the state of expression of a gene that does not involve a mutation, but that is nevertheless inherited in the absence of the signal (or event) that initiated that change." Many of the regulatory events we have discussed are epigenetic. For example, referring back to the λ case in bacteria, transient expression of one protein (CII) activates transcription of a gene (*cI*), the product of which then keeps that gene on, a state that is inherited.

Another example is found with the *lac* genes. A low level of lactose in the medium does not activate the *lac* genes in ordinary cells. Cells that are preinduced with a pulse of lactose at higher concentration will maintain that induction if the lactose level is then reduced (to the low level in question). The effect is caused by induction of a permease (by the initial, high, levels of lactose) that concentrates lactose in the cell. Once permease is produced, it can maintain a high level of lactose in the cell even when the concentration of that sugar in the medium is low.

Many of the gene regulatory events underlying development of a higher organism involve these kinds of epigenetic changes. In contrast, there is an array of epigenetic phenomena in eukaryotes—particularly in higher eukaryotes and in the yeast *Schizosaccharomyces pombe*—that evidently cannot be explained solely by the kinds of mechanisms alluded to above. The key difference is that in these cases, there is a different effect on one of two homologous genes or chromosomes in the same cell—a so-called *cis* effect on gene expression. The classic example is X-chromsome inactivation in mammals, in which the genes on one of the two X chromosomes in each cell is, at an early stage of development, largely inactivated, and this state is stably maintained throughout development.

Our understanding of mechanisms underlying such *cis*-epigenetic effects is rudimentary. It has been suggested that in at least some such cases, histone modifications that can help or hinder gene expression can, once established, be inherited.

We have mentioned that acetylated histones are recognized by so-called bromo domains found in several components of the transcriptional machinery, including HATs. A similar picture describes histone methylation: enzymes (found in higher eukaryotes and in the yeast *S. pombe*) methylate specific lysines in histone tails; those modified nucleosomes are recognized by proteins bearing chromo domains, including proteins that are themselves histone methylases. Phosphorylation of histones provides another means of creating such a "histone code."

The difficulty with such models is in understanding how the effect would be transmitted upon replication; i.e., the modified histones would have to be inherited (and perpetuated) only on the chromosomes that are the descendants of the one originally modified. We do not know whether such a scenario holds in any specific case. See also Footnote 19.

state were formed, it would tend to be self-perpetuating because the Sir proteins, some of which are HDACs, themselves have higher affinity for unacetylated histones.

Silencing at the HM loci is stronger than that at the telomere, but there too, a strong activator overcomes the effect. Variegation is not ordinarily observed with a gene placed at these loci, but it is observed if one of the relevant Sir proteins is mutated.[42]

Variegation is often cited as an example of "epigenetic" change.

DNA LOOPING

We introduced our discussion of silencing by emphasizing how moving a gene to heterochromatic regions of the yeast genome can diminish that gene's expression. In this section, we describe a situation in which, seemingly paradoxically, moving a gene to such a region *increases* its expression. But we can explain the result from what we have already learned; moreover, the result shows how an activator, bound some distance from a gene, can nevertheless recruit the transcriptional machinery to that gene.

We have noted that the typical yeast activator works when positioned some 250 bp upstream of a gene. The effect of activators in yeast typically diminishes as the activator-binding sites are moved away from the gene. For example, Gal4 activates weakly but detectably when the UAS_G is positioned 500 bp upstream of the gene, but when positioned more than 1000 bp away, virtually no activation is observed. Likewise, Gal4 has no effect when positioned downstream from—and hence more than 1000 bp from the start of—a typical gene.

Our discussion of cooperative binding of proteins in Chapter 1 provides a likely explanation: as sites are moved further apart on a DNA molecule, it becomes increasingly difficult for proteins, e.g., activator and transcriptional machinery, to bind cooperatively to them. Cooperative binding would, according to the simplest scenario, require that the DNA between the sites loop out to accommodate the reaction.[43]

If looping is indeed the mechanism that allows communication between the DNA-bound activator and the transcribing machinery at the promoter, anything that helps loop formation should increase the distance from which an activator can work. An explicit demonstration is provided

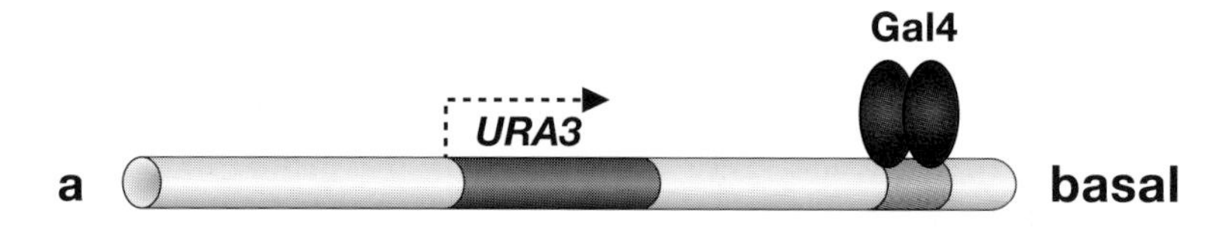

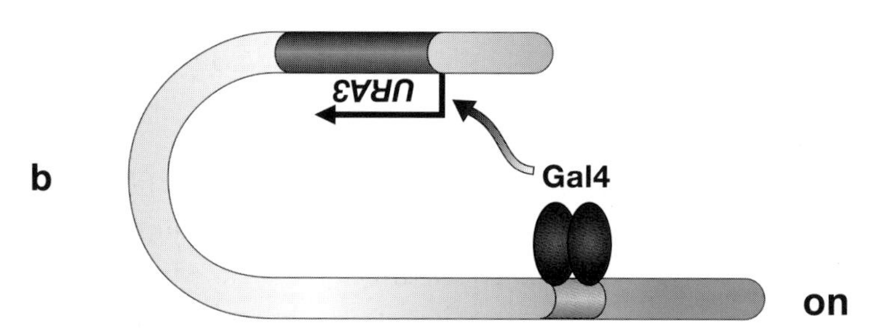

FIGURE 2.21. Activation "at a distance." An activator bound downstream from the gene has no effect if the construct is at an internal location on the chromosome (*a*). When placed at the telomere (*b*), however, the activator works from downstream. In this case, the folded heterochromatin facilitates activation.

by an experiment in which a gene bearing an activator-binding site far from the gene was placed near a telomere.

In this case, Gal4 activated the gene *URA3* from a construct in which the Gal4 sites were placed more than 1000 bp downstream from the promoter; but activation was only seen when that construct was placed near a telomere (Figure 2.21). Evidently, the "built in" loop of the telomeric heterochromatin brings the UAS_G close to the gene so that Gal4 can recruit the transcriptional machinery to the promoter. If the activator communicated with the gene by some mechanism other than by looping—by moving along the DNA, for example —the imposed loop would be expected to have no effect.[44]

These results might help to explain a curious effect, long noted in the study of *Drosophila* genes. In *Drosophila*, as in yeast, most genes show diminished activity when positioned in heterochromatin, but a few are known that function better when positioned there. Perhaps these genes, like the yeast constructs of Figure 2.21, require preformed loops for efficient activation, loops that would be provided by the *Drosophila* heterochromatin.

SUMMARY

In bacteria, certain activators work by recruiting RNA polymerase to one or another gene. In yeast, the only activators we know of also work by recruiting polymerase, but they do so in the face of a vastly more complicated transcriptional machinery. This machinery includes, in addition to polymerase, complexes involved directly in transcribing the gene and nucleosome modifiers that can affect access of the gene.

In yeast, new functions have been accommodated in this regulatory scheme. For example, where nucleosome modifiers are required, the same activating region that recruits proteins directly associated with polymerase recruits these enzymes as well. Where new demands are imposed—as at different stages of the cell cycle or a change in gene location—the existing activation mechanism continues to work. As we shall also see in Chapter 3, where a factor is required for elongation, that too will be recruited by an activator. Thus, Gal4 will work on virtually any gene, under any physiological condition, by recruiting whatever components are required.

Moreover, hints we saw in bacteria are clearly illustrated in yeast: recruitment readily lends itself to combinatorial control and to signal integration. We will also see these themes elaborated further in Chapter 3.

FOOTNOTES

[1]Yeast, like all eukaryotes, have two additional RNA polymerases. PolI transcribes genes encoding RNA that functions as part of ribosomes; PolIII transcribes mainly tRNA and related genes. Genes encoding proteins are, with few exceptions, transcribed by PolII.

[2]The elongating polymerase is phosphorylated, especially on the carboxy-teminal domain (CTD) or "tail" of its largest subunit. In Chapters 3 and 4, we will encounter a kinase called P-TEFb that, in higher eukaryotes, promotes elongation by phosphorylating the polymerase. Yeast are suspected to contain one or more kinases with similar function. It is not clear what role the kinase in TFIIH (called Cdk7) might have in this regard, but the helicases are known to perform essential functions at the early stages of transcription. The phosphorylated tail of polymerase binds enzymes needed for mRNA processing. TFIIH also has an important role in repairing damaged DNA.

[3]As implied, these proteins are found in various states of aggregation depending on the conditions and the methods used by the investigator. For example, there are apparently two major forms of the mediator, one of which includes the "Srb8-11" subcomplex and one of which does not. The latter subcomplex is depeleted in cells grown to saturation. There is disagreement as to the physiological significance of the holoenzyme as a single complex. Estimates of the percentage of polymerase bound to the mediator, when not on DNA, range from very low (less than a few percent) to quite high (more than 50%).

[4]Four genes are required for galactose metabolism: *GAL1, GAL2, GAL7*, and *GAL10*. A UAS_G lying between the divergently transcribed *GAL1* and *GAL10* genes controls both; separate UAS_Gs control *GAL2* and *GAL7*. Gal4 also activates three other genes: *GAL80* and *GAL3*, the products of which are involved in regulating induction, and *MEL1*, the product of which generates galactose and glucose from the sugar mellibiose. Each of these three genes is activated by Gal4 working at a single DNA-binding site.

[5]Footnote 4 referred to three genes (*GAL80, GAL3*, and *MEL1*) that are activated from single Gal4-binding sites. These genes are less tightly regulated by Gal80 than are those genes activated from a UAS_G. That is, in the absence of galactose, there is a low but significant level of expression of these genes. The explanation, evidently, is that interaction between Gal80 molecules bound to adjacent Gal4 molecules increases the efficiency of inhibition. For two of these genes, this scenario makes obvious sense: *GAL80* must be present in the absence of galactose to keep Gal4 silent, and *GAL3* must be present to allow induction.

[6]Galactose does not completely relieve Gal80 inhibition of Gal4. Thus, Gal4 works approximately two- to threefold more efficiently in a strain deleted for Gal80 than in a wild-type cell growing in galactose.

[7]The kinase that works on Mig1 is Snf1. Phosphorylated Mig1 binds a protein called a karyopherin (in this case Msn5) which carries Mig1 out of the nucleus.

[8]Activating regions work when tethered to DNA artificially. Attaching an activating region to a synthetic polyamide, designed to recognize a specific base sequence, activates transcription in assays performed with yeast extracts.

[9]Various experiments suggest that the typical activating region has no defined secondary structure but that such activating regions form α-helices upon interaction with their targets. Gcn4, Hap4, and Ppr1 are examples of other yeast acidic activators.

[10]Similar observations have been made with another acidic yeast activator, Gcn4; in that case, seven hydrophobic patches have been identified as contributing to the activation function.

[11]Consistent with this surmise, substitution of various amino acids in ARII with proline—a residue that typically disrupts α-helices and β-sheets—abolishes interaction with Gal80, but again has a negligible effect on activation.

[12]Such peptides can be found encoded in random bits of DNA, as in Figure 2.9 or can be generated by synthesis. For example, a 15-residue peptide, AH, designed to form an amphipathic helix with acidic, hydrophobic, and polar surfaces, works as an activating region. There is no evidence that this peptide actually forms an α-helix. An interesting exception to the general rule that acidic residues are essential is XL. This synthetic activating region, eight residues long, is nonacidic, and yet works as an efficient activator. It apparently sees a target (or targets) other than those recognized by the acidic activators.

[13]Overexpression of Gal4 in wild-type cells grown in the absence of galactose can activate the *GAL* genes. This happens because, at moderately high levels, Gal4 titrates Gal80, thereby relieving inhibition of the DNA-bound Gal4. At still higher levels of Gal4, however, squelching is observed.

[14]The name Gal11 is a historical accident; the protein has no special role at the *GAL* genes. Its elimination (by deletion) decreases transcription of some 50 genes and modestly increases transcription of approximately 85 others. Gal11 does not itself bind DNA, and the P mutation in Gal11P has no known effect in the absence of a DNA-bound Gal4 fragment.

[15]Mutations in the region of Gal4 that binds Gal11P (i.e., residues 65–100) alter the strength of the interaction with Gal11P and affect activation accordingly. The dissociation constant describing the reaction between Gal4(1–100) and the original Gal11P mutant is approximately equivalent to that measured for the interaction between a typical acidic activating region and a putative target, TBP, i.e., about 10^{-7} M.

[16]Where a given fusion protein does not elicit significant levels of transcription (Lex-Srb4 working at *GAL1,* for example), the following experiment indicates that the fusion is nevertheless recruiting at least part of the transcriptional machinery: when bound to DNA along with a weak natural activator, a large synergistic effect is observed. At least anecdotally, many fusion proteins (e.g., Lex-Srb7) elicit more transcription from reporters carried on plasmids than from chromosomally integrated reporters. Whether this is a consequence of increased copy number of the reporter, and/or to a less restrictive chromatin configuration on the plasmid, is not known. Gene activation elicited in bypass experiments does not mimic in all ways that elicited by natural activators. Thus, for example, the former is more efficiently repressed by DNA-tethered Tup1 than the latter. The reason for this difference is not understood.

[17]In a further experiment, activating regions were fused to various components of the transcriptional machinery (including Gal11, Srb2, TBP, TAF17, TAF23). In no case

was spontaneous transcription of a reporter observed. The result is consistent with the notion that activators work by recruitment.

[18]It is often assumed that nucleosomes repress transcription in yeast, but it has been difficult to test this idea. It is possible to deplete cells (specially constructed for this purpose) of a single histone and to assay gene expression in the dying cells. In such cells, it is found that about 10% of genes show an increase in transcription and about 10% a decrease, whereas the majority remain unchanged.

[19]Full expression of the *Ino1* gene is reported to require both acetylation and phosphorylation of nucleosomes associated with the gene. Both modifications occur on histone H-3: Phosphorylation at Ser-10 is required for the subsequent acetylation of Lys-9. Presumably, the recruited kinase (Snf1) modifies the nucleosome so as to increase its affinity for the HAT (Gcn5).

[20]Certain mammalian activators and other DNA-binding proteins get acetylated, for example, modifications that can affect their activities. Histones are notoriously good substrates for enzymes, even some whose action on them is not believed to be of any physiological importance. For example, cyclin-dependent kinases readily phosphorylate histones in vitro, a reaction that is not believed to be physiologically meaningful.

[21]In this experiment, nucleosome modifiers, as well as the transcribing machinery, were required for transcription. The template was first modified as a result of recruitment of the modifiers by the activator. The modified template was separated from the modifying enzymes, and the transcribing machinery was added. Thus, under these conditions, nucleosome modification, while required, was not sufficient to elicit significant levels of transcription. This result mirrors the following observation made in vivo: artificially tethering HATs to DNA (e.g., using LexA-Gcn5 fusion) typically activates transcription weakly, if at all.

[22]As would be expected for a system driven by recruitment, the results of experiments performed in vitro can depend on the concentrations of the reactants. For example, it has been reported that an activator is not required for high levels of transcription at high concentrations of polymerase and associated factors. Similarly, the requirement for an activator to recruit a HAT can be obviated by increasing the concentration of that HAT. It follows that a component required for transcription in a given case—a nucleosome modifier, for example—might or might not be present in the cell at sufficiently high concentrations to work as part of the background; i.e., it might or might not have to be recruited. Another problem arises from the large number of proteins that can be involved in transcribing a gene. Thus, for example, many experiments with relatively purified components demonstrated promoter-specific transcription in the absence of components known to be required in vivo, Srb4, for example. Only recently has transcription, and transcriptional activation, been reproduced in the test tube in a fashion that requires many of the components known to be required in vivo.

[23]The identification of any given protein or modification depends on the availability of an antibody specific for that protein or modified protein. But even then there can be problems. For example, sometimes the epitope recognized by the antibody is covered by other proteins and so no signal is seen. In some experiments, proteins are

tagged with epitopes for which antibodies are readily available; such manipulations can affect the results in unexpected ways. Should one or more of the bound components become modified—acetylated, for example—that modification might hide or reveal an epitope. The problem of interpreting the significance of small differences (e.g., factors of 2–3) is particularly acute.

[24]It may be impossible to distinguish indirect from direct effects on expression of any given gene. And, as these experiments are usually performed, one can at best detect relative changes. For example, if all genes are effected equally, it might appear that none was affected.

[25]One problem here is being confident that all traces of the protein in question are eliminated, and, of course, there could be indirect effects of dying.

[26]There is disagreement as to whether the level of acetylation of H3 tails in nucleosomes at the *GAL1* promoter goes up upon induction in wild-type cells.

[27]The genes affected in this way by deletion of Snf/Swi or Gcn5 comprise overlapping sets. Another 5% of genes are expressed at higher levels in such deletion strains. Deletion of both Swi/Snf and Gcn5 is lethal.

[28]The first of these experiments was performed with the *GAL1* gene at its ordinary location. If a sequence known to bind and position a nucleosome is introduced adjacent to the two weak sites in this experiment, the requirement for the modifiers is relieved. The interpretation is that DNA immediately adjacent to the positioned nucleosome is histone-free, and thus the modifiers in this case are not required. Such strong nucleosome-positioning sequences are rare, and it is not clear what are the sequence requirements to form such a site.

[29]How then do cells lacking one or another of those modifiers survive? The answer seems to be that these "mitosis" genes can be transcribed in the absence of the modifiers, but only at a somewhat later stage of the cell cycle. The cells are sick, presumably because of this delay, but they do survive.

[30]It is reported that deletion of this inhibitory flap, while having little effect on activation by Gal4, rather dramatically improves activation by fusion proteins bearing DNA-binding domains attached to Gal11 and Srb4. The result suggests that indeed acidic activators have a way of overcoming the inhibitory flap that the artificial activators do not have.

[31]It is reported that Srb10, the kinase in the mediator, is contacted by acidic activating regions as they work. That contact, in turn, is believed to promote phosphorylation of activators outside the activating regions with different consequences in different cases. Thus, following phosphorylation, the activator Gcn4 is proteolyzed and the activator Msn2 is exported from the nucleus. Conceivably, this activator destruction could enable the activator to work stepwise. Gal4 is also phosphorylated by Srb10, a modification that helps it overcome the inhibitory effect of Gal80. Srb10 also has a role in mediating the effects of repressors (see text).

[32]We have only touched upon the question of whether different activators might touch different targets (see Footnote 12). It has been suggested that certain activators recruit polymerase by a route different from that used by Gal4. Ace1, for example, unlike Gal4, is reported to activate independently of either the mediator com-

ponent Srb4 or of the helicases in TFIIH. It is suggested that Ace1 recruits a form of the mediator, lacking the Srb proteins, that cannot be recruited by Gal4. To what extent the composition of the mediator might vary in vivo (see Footnote 3) and the extent to which this might influence activation by particular activators is not clear.

[33]In brief, yeast grow as haploids of two types—**a** and α—and, following fusion of these two cell types, as diploids. Protein **a**1 is made in haploids of type **a** as well as in diploids; α1 and α2 are made in α cells and in diploids; and Mcm1 is made ubiquitously. Thus, **a**1 and α2 are found together only in the diploid, where they repress "haploid-specific"genes. The combination Mcm1/α1 activates so-called α-specific genes, and, in the same cell, the combination Mcm1/α2 represses so-called **a**-specific genes.

Haploid **a** and α cells switch to the opposite form about once per generation. This switch, which occurs only in mother cells (see later in the text), involves a DNA rearrangement that brings previously silent regulatory genes under control of active promoters. **a**1 and α2 are two of these regulators, and that is why their expression is cell-type-specific.

[34]Exactly why Swi5 can bind without help, whereas SBF cannot, is not known. Presumably the answer lies in the differences in the strengths with which the activators bind and/or the disposition of nucleosomes, and so on.

[35]Recruitment of the mediator can be separated from recruitment of the polymerase by SBF as follows. The kinase Cdk1 is required for the transition from the G_1 to the S phase of the cell cycle. In cells bearing a temperature-sensitive Cdk1 and transferred to the nonpermissive temperature, SBF is found to recruit the mediator but not the polymerase. Polymerase recruitment follows upon lowering the temperature of these blocked cells. This effect of Cdk1 does not apply to the *GAL* genes, where Gal4 recruits the mediator and polymerase indistinguishably, even in the absence of active Cdk1. How Cdk1 exerts this effect at the *HO* promoter is not understood.

[36]The biological importance of this form of signal integration is to restrict expression of *HO* to one stage of the cell cycle but only in so-called "mother" cells. A yeast cell divides by budding: a mother cell produces a bud, which becomes a daughter cell. Swi5 moves to the nucleus only at a specific point in the cell cycle, but does so in both mother and daughter cells. Restriction of expression to the mother cell is determined by a repressor called Ash1. Ash1 protein is made only in daughter cells. Ash1 binds to sites in the *HO* promoter and prevents activation by Swi5. The mRNA encoding Ash1, made in the mother cell, is transported to the developing daughter bud before translation. It, as well as the *Ash1* mRNA made by the daughter, is translated into protein specifically in the daughter cell. Cell-cycle-specific phosphorylation also controls the activity of SBF.

[37]The ordinary role of the silencing regions at the telomere and the ribosomal DNA loci is not to silence genes. There are no genes at the telomere (normally), and relieving silencing at the rDNA locus has no effect on gene expression; strong activators that ordinarily work there overcome any silencing effect. Rather, the silencing structure found at telomeres helps to stabilize telomeres. The silencing structure at the rDNA locus blocks unwanted recombination between the multiple gene copies. At the HM loci, genes that determine mating type are silenced, but even in that case,

there is another crucial role: to prevent recombination with the so-called active mating-type locus.

[38]Many yeast genes, at their ordinary chromosomal locations, express a physiologically significant level of product even in the absence of the activator that induces them in response to the appropriate environmental conditions. In some cases, it is known that constitutively present proteins bind to sites near the promoters of these genes, and these proteins contribute to this "basal" expression. Whether such proteins are always required for significant basal expression is not known. In many cases, the defined activators only increase expression approximately two- to threefold.

[39]In higher eukaryotes, heterochromatin, unlike euchromatin, is highly condensed throughout the cell cycle. Heterochromatin bears characteristic proteins not found in euchromatin. DNA in heterochromatin is less accessible to DNA cutting enzymes, for example.

[40]Sir2 is an unusual HDAC in that it requires the cofactor nicotinamide adenine dinucleotide. Loss of silencing has been suggested as a cause of concomitant aging in yeast, and it has been suggested that this cofactor tends to be depleted in older cells. Sir2 is also a ribosylase, and it is currently thought that self-ribosylation activates the HDAC activity.

[41]Telomeres may be a sink for other proteins as well. Thus, induction of double-strand breaks anywhere in the yeast genome causes release of the Ku70/80 heterodimer—another telomere component—from the telomere, and the protein moves to the site of breakage to mediate DNA repair.

[42]It has been suggested that at least part of the transcriptional machinery can be recruited to a gene within HM by a weak activator, without evident transcription. Presumably, the stronger activator recruits missing components. We return to an explicit example of such a phenomenon in Chapter 3.

[43]As the sites are moved farther apart, the entropic costs of those sites coming together increases. What is rather more surprising is the ability of activators similar to Gal4 to work over very large distances in higher eukaryotes, a matter we return to in Chapter 3.

[44]The effect depends on the *SIR3* gene, which in turn is required for formation of the telomeric folded structure. The effect is observed with the promoter-UAS_G construct in either orientation with respect to the end of the chromosome.

BIBLIOGRAPHY

Cold Spring Harbor Symposia on Quantitative Biology. 1998. Volume 63: Mechanisms of transcription. Cold Spring Harbor Laboratory Press, Cold Spring Harbor, New York.

Latchman D.S. 1998. *Eukaryotic transcription factors*, 3rd edition. Academic Press, London.

Russo V.E.A., Martienssen R.A., and Riggs A.D., eds. 1996. *Epigenetic mechanisms of gene regulation.* Cold Spring Harbor Laboratory Press, Cold Spring Harbor, New York.

Strathern J.N., Jones E.W., and Broach J.R., eds. 1982. *The molecular biology of the yeast* Saccharomyces: *Metabolism and gene expression.* Cold Spring Harbor Laboratory, Cold Spring Harbor, New York.

White R.J. 2001. *Gene transcription: Mechanisms and control.* Blackwell Science, Malden, Massachusetts.

RNA Polymerase and the Transcriptional Machinery

Bentley D.L. 1995. Regulation of transcriptional elongation by RNA polymerase II. *Curr. Opin. Genet. Dev.* **5:** 210–216.

Buratowski S. 2000. Snapshots of RNA polymerase II transcription initiation. *Curr. Opin. Cell Biol.* **12:** 320–325.

Burley S.K. and Roeder R.G. 1998. TATA box mimicry by TFIID: Autoinhibition of pol II transcription. *Cell* **94:** 551–553.

Dvir A., Conaway J.W., and Conaway R.C. 2001. Mechanism of transcription initiation and promoter escape by RNA polymerase II. *Curr. Opin. Genet. Dev.* **11:** 209–214.

Green M.R. 2000. TBP-associated factors (TAFIIs): Multiple, selective transcriptional mediators in common complexes. *Trends Biochem. Sci.* **25:** 59–63.

Hampsey M. 1998. Molecular genetics of the RNA polymerase II general transcriptional machinery. *Microbiol. Mol. Biol. Rev.* **62:** 465–503.

Lee T.I. and Young R.A. 1998. Regulation of gene expression by TBP-associated proteins. *Genes Dev.* **12:** 1398–1408.

Lee T.I. and Young R.A. 2000. Transcription of eukaryotic protein-coding genes. *Annu. Rev. Genet.* **34:** 77–137.

Malik S. and Roeder R.G. 2000. Transcriptional regulation through mediator-like coactivators in yeast and metazoan cells. *Trends Biochem. Sci.* **25:** 277–283.

Myers L.C. and Kornberg R.D. 2000. Mediator of transcriptional regulation. *Annu. Rev. Biochem.* **69:** 729–749.

Orphanides G. and Reinberg D. 2000. RNA polymerase II elongation through chromatin. *Nature* **407:** 471–475.

Orphanides G., Lagrange T., and Reinberg D. 1996. The general transcription factors of RNA polymerase II. *Genes Dev.* **10:** 2657–2683.

Parvin J.D. and Young R.A. 1998. Regulatory targets in the RNA polymerase II holoenzyme. *Curr. Opin. Genet. Dev.* **8:** 565–570.

Struhl K. 1997. Selective roles for TATA-binding-protein-associated factors in vivo. *Genes Funct.* **1:** 5–9.

Winston F. 2001. Control of eukaryotic transcription elongation. *Genome Biol.* **2:** REVIEWS1006.

Nucleosome Modifiers

Bash R. and Lohr D. 2001. Yeast chromatin structure and regulation of *GAL* gene expression. *Prog. Nucleic Acid Res.* **65:** 197–259.

Berger S.L. 1999. Gene activation by histone and factor acetyltransferases. *Curr. Opin. Cell Biol.* **11:** 336–341.

Brown C.E., Lechner T., Howe L., and Workman J.L. 2000. The many HATs of transcriptional coactivators. *Trends Biochem. Sci.* **25:** 15–19.

Cavalli G. and Paro R. 1998. Chromo-domain proteins: Linking chromatin structure to epigenetic regulation. *Curr. Opin. Cell Biol.* **10:** 354–360.

Flaus A. and Owen-Hughes T. 2001. Mechanisms for ATP-dependent chromatin remodeling. *Curr. Opin. Genet. Dev.* **11:** 148–154.

Grunstein M. 1997. Histone acetylation in chromatin structure and transcription. *Nature* **389:** 349–352.

Jenuwein T. and Allis C.D. 2001. Translating the histone code. *Science* **293:** 1074–1080.

Kingston R.E. and Narlikar G.J. 1999. ATP-dependent remodeling and acetylation as regulators of chromatin fluidity. *Genes Dev.* **13:** 2339–2352.

Kornberg R.D. and Lorch Y. 1999. Twenty-five years of the nucleosome, fundamental particle of the eukaryote chromosome. *Cell* **98:** 285–294.

Kornberg R.D. and Lorch Y. 1999. Chromatin-modifying and -remodeling complexes. *Curr. Opin. Genet. Dev.* **9:** 148–151.

Marmorstein R. and Roth S.Y. 2001. Histone acetyltransferases: Function, structure, and catalysis. *Curr. Opin. Genet. Dev.* **11:** 155–161.

Peterson C.L. and Workman J.L. 2000. Promoter targeting and chromatin remodeling by the SWI/SNF complex. *Curr. Opin. Genet. Dev.* **10:** 187–192.

Rice J.C. and Allis C.D. 2001. Histone methylation versus histone acetylation: New insights into epigenetic regulation. *Curr. Opin. Cell Biol.* **13:** 263–273.

Roth S.Y., Denu J.M., and Allis C.D. 2001. Histone acetyltransferases. *Annu. Rev. Biochem.* **70:** 81–120.

Struhl K. 1998. Histone acetylation and transcriptional regulatory mechanisms. *Genes Dev.* **12:** 599–606.

Sudarsanam P. and Winston F. 2000. The Swi/Snf family nucleosome-remodeling complexes and transcriptional control. *Trends Genet.* **16:** 345–351.

Vignali M., Hassan A.H., Neely K.E., and Workman J. 2000. ATP-dependent chromatin-remodeling complexes. *Mol. Cell. Biol.* **20:** 1899–1910.

Widom J. 1998. Structure, dynamics, and function of chromatin in vitro. *Annu. Rev. Biophys. Biomol. Struct.* **27:** 285–327.

Widom J. 2002. Role of DNA sequence in nucleosome stability and dynamics. *Q. Rev. Biophys.* (in press).

Winston F. and Allis C.D. 1999. The bromodomain: A chromatin-targeting module? *Nat. Struct. Biol.* **6:** 601–604.

Activation and Repression

Barberis A. and Gaudreau L. 1998. Recruitment of the RNA polymerase II holoenzyme and its implications in gene regulation. *Biol. Chem.* **379:** 1397–1405.

Berk A.J. 1999. Activation of RNA polymerase II transcription. *Curr. Opin. Cell Biol.* **11:** 330–335.

Carlson M. 1999. Glucose repression in yeast. *Curr. Opin. Microbiol.* **2:** 202–207.

Fry C.J. and Peterson C.L. 2001. Chromatin remodeling enzymes: Who's on first? *Curr. Biol.* **11:** R185–197.

Hahn S. 1998. Activation and the role of reinitiation in the control of transcription by RNA polymerase II. *Cold Spring Harbor Symp. Quant. Biol.* **63:** 181–188.

Johnson A.D. 1995. Molecular mechanisms of cell-type determination in budding yeast. *Curr. Opin. Genet. Dev.* **5:** 552–558.

Johnston M. 1987. A model fungal gene regulatory mechanism: The *GAL* genes of *S. cerevisiae. Microbiol. Rev.* **51:** 458–476.

Johnston M. 1999. Feasting, fasting and fermenting: Glucose sensing in yeast and other cells. *Trends Genet.* **15:** 29–33.

Jones K.A. and Kadonaga J.T. 2000. Exploring the transcription-chromatin interface. *Genes Dev.* **14:** 1992–1996.

Kornberg R.D. 1999. Eukaryotic transcriptional control. *Trends Cell Biol.* **9:** M46–49.

Maldonado E., Hampsey M., and Reinberg D. 1999. Repression: Targeting the heart of the matter. *Cell* **99:** 455–458.

Melcher K. 1997. Galactose metabolism in *S. cerevisiae:* A paradigm for eukaryotic gene regulation. In *Yeast sugar metabolism* (ed. F.K. Zimmermann and K.-D. Entian), pp. 235–269. Technomic Publishing, Lancaster, Pennsylvania.

Naar A.M., Lemon B.D., and Tjian R. 2001. Transcriptional coactivator complexes. *Annu. Rev. Biochem.* **70:** 475–501.

Oshima Y. 1982. Regulatory circuits for gene expression: The metabolism of galactose and phosphate. In *Molecular biology of the yeast* Saccharomyces: *Metabolism and gene expression* (ed. J.N. Strathern et al.), pp. 169–180. Cold Spring Harbor Laboratory, Cold Spring Harbor, New York.

Ptashne M. 1988. How eukaryotic transcriptional activators work. *Nature* **335:** 683–689.

Ptashne M. and Gann A. 1997. Transcriptional activation by recruitment. *Nature* **386:** 569–577.

Smith R.L. and Johnson A.D. 2000. Turning genes off by Ssn6-Tup1: A conserved system of transcriptional repression in eukaryotes. *Trends Biochem. Sci.* **25:** 325–330.

Struhl K. 1995. Yeast transcriptional regulatory mechanisms. *Annu. Rev. Genet.* **29:** 651–674.

Struhl K. 1996. Chromatin structure and RNA polymerase II connection: Implications for transcription. *Cell* **84:** 179–182.

Struhl K. 1999. Fundamentally different logic of gene regulation in eukaryotes and prokaryotes. *Cell* **98:** 1–4.

Tansey W.P. 2001. Transcriptional activation: Risky business. *Genes Dev.* **15:** 1045–1050.

Triezenberg S.J. 1995. Structure and function of transcriptional activation domains. *Curr. Opin. Genet. Dev.* **5:** 190–196.

Wolberger C. 1998. Combinatorial transcription factors. *Curr. Opin. Genet. Dev.* **8:** 552–559.

Silencing and Variegation

Bi X. and Broach J.R. 2001. Chromosomal boundaries in *S. cerevisiae. Curr. Opin. Genet. Dev.* **11:** 199–204.

Cockell M. and Gasser S.M. 1999. Nuclear compartments and gene regulation. *Curr. Opin. Genet. Dev.* **9:** 199–205.

Defossez P.A., Lin S.J., and McNabb D.S. 2001. Sound silencing: The Sir2 protein and cellular senescence. *Bioessays* **23:** 327–332.

Gartenberg M.R. 2000. The Sir proteins of *Saccharomyces cerevisiae:* Mediators of transcriptional silencing and much more. *Curr. Opin. Microbiol.* **3:** 132–137.

Gasser S.M. 2001. Positions of potential: Nuclear organization and gene expression. *Cell* **104:** 639–642.

Gottschling D.E. 2000. Gene silencing: Two faces of SIR2. *Curr. Biol.* **10:** R708–711.

Grewal S.I. 2000. Transcriptional silencing in fission yeast. *J. Cell. Physiol.* **184:** 311–318.

Grunstein M. 1998. Yeast heterochromatin: Regulation of its assembly and inheritance by histones. *Cell* **93:** 325–328.

Guarente L. 2001. SIR2 and aging—The exception that proves the rule. *Trends Genet.* **17:** 391–392.

Lustig A.J. 1998. Mechanism of silencing in *S. cerevisiae. Curr. Opin. Genet. Dev.* **8:** 233–239.

Moazed D. 2001. Enzymatic activities of Sir2 and chromatin silencing. *Curr. Opin Cell Biol.* **13:** 232–238.

Shore D. 2001. Telomeric chromatin: Replicating and wrapping up chromosome ends. *Curr. Opin. Genet. Dev.* **11:** 189–198.

Two-hybrid Assay

Brent R. and Finley R.L., Jr. 1997. Understanding gene and allele function with two-hybrid methods. *Annu. Rev. Genet.* **31:** 663–704.

Fields S. and Sternglanz R. 1994. The two-hybrid system: An assay for protein-protein interactions. *Trends Genet.* **10:** 286–292.

Hazbun T.R. and Fields S. 2001. Networking proteins in yeast. *Proc. Natl. Acad. Sci.* **98:** 4277–4278.

Topogenic Sequences

Pemberton L.F., Blobel G., and Rosenblum J.S. 1998. Transport routes through the nuclear pore complex. *Curr. Opin. Cell Biol.* **10:** 392–399.

CHAPTER 3

Some Notes on Higher Eukaryotes

Muddy water is not necessarily deep.

FRIEDRICH NIETZSCHE

HIGHER EUKARYOTES HAVE MORE GENES THAN YEAST: by current estimates, 15,000 for a fly and approximately 30,000 for a human. In addition, they engage in more RNA splicing, thereby generating more messenger RNAs from the same number of genes. What we are most concerned with, however, are the elaborate transcriptional controls placed on many genes of higher eukaryotes, controls that enable a given gene to be expressed at many different times and places during development.

To achieve this, three strategies of transcriptional regulation are exploited to a greater extent in higher eukaryotes than in yeast.

- ***There is more signal integration:*** For many genes, activation depends on the simultaneous presence of multiple activators, each of which might be controlled by a separate signal.
- ***There is more combinatorial control:*** Many activators work in various combinations with other activators (and repressors) to regulate different sets of genes.
- ***Genes are often controlled by alternative sets of regulators:*** This allows the same gene to be expressed in response to different sets of signals.

As one would expect from this description, the typical gene in higher eukaryotes is associated with a greater number of regulatory protein-binding sites than is its yeast counterpart. Those sites are often dispersed over thousands of base pairs, and so the regulators must often work "at a distance." A set of contiguous regulatory binding sites is called an enhancer, and enhancers are sometimes found to work over many thousands of base pairs upstream and downstream from genes.

MECHANISM OF ACTIVATION: RECRUITMENT

In keeping with the approach adopted thus far in this book, our first task is to decipher how the typical transcriptional activator works in higher eukaryotes. Recall that in bacteria, we encountered three mechanisms—polymerase activation, promoter activation, and regulated recruitment—and that of these three, regulated recruitment best describes how the typical yeast activator works. It is more difficult to perform genetic and biochemical experiments with higher eukaryotes than with bacteria and yeast, and so in attempting to asnswer this first question, we are guided by what we learned in the previous chapters. Here are some salient observations:

- The yeast activator Gal4 can activate genes in higher eukaryotes. The protein, expressed in cells of these organisms, activates genes to which Gal4-binding sites have been added. This experiment has been performed with genes in mammalian, plant, and insect cells. Gal4 is routinely used to express genes artificially in *Drosophila*; as far as we know, Gal4 will activate any gene bearing Gal4 sites, in any tissue of the fly (see Figure 3.1).

- Many activators in higher eukaryotes resemble yeast activators in that their activating regions can be separated from their DNA-binding domains. These activating regions work when attached to heterologous DNA-binding domains. This result is inconsistent with a mechanism in which the nature of DNA binding is inextricably linked to activation, as we found was the case for the bacterial activator MerR. For an example in which DNA binding has an allosteric effect on the activating region, see Footnote 1.

- Many higher eukaryotic activating regions, particularly strong ones like their yeast counterparts, bear an excess of hydrophobic and acidic residues. Such regions are found, for example, on the mammalian proteins VP16 (encoded by a herpesvirus), p53, and NFκB, and the *Drosophila* protein Bicoid. There are non-acidic activating regions as well. For example, a so-called glutamine-rich activating region is found

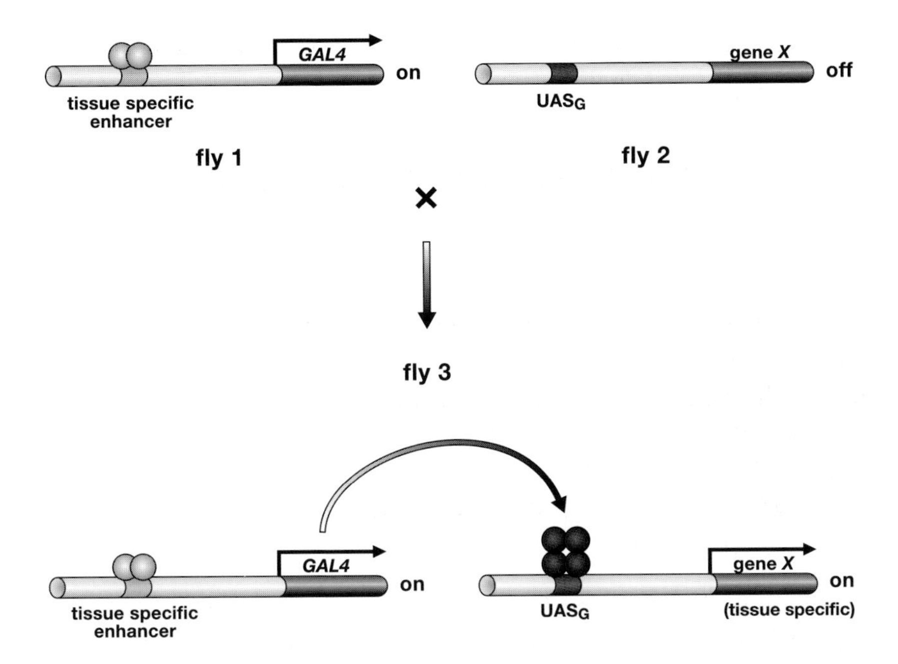

FIGURE 3.1. Tissue-specific expression of *Drosophila* genes using Gal4. The construct on the left bears a *GAL4* gene attached to a promoter that drives transcription in specific tissues; it is introduced into one fly line. The construct on the right bears a gene *X*, attached to a UAS_G; it is introduced into a second fly line. When the lines are interbred, progeny contain both constructs. In such flies, gene *X* is expressed wherever Gal4 is present. Because there are many fly lines that express Gal4 in different specific tissues, any gene—placed under control of a UAS_G—can be expressed in a wide array of specified and different patterns.

on the mammalian activator SP1. Several of these activating regions have other characteristics similar to their yeast counterparts. For example, among activating region deletion mutants, activation is found to be roughly proportional to length. This and other properties are, as we argued in Chapter 2, suggestive of adhesive surfaces.[2]

- As in yeast, activating regions must be tethered to DNA to activate transcription. Overproduction of activating regions, lacking DNA-binding domains, does not result in activation. At sufficiently high concentration, especially with strong activating regions, overproduction causes squelching (see Chapter 2). These results are inconsistent with the "polymerase activation" model—in that case, as was seen for the bacterial activator NtrC, overexpression in the absence of DNA binding leads to activation of a prebound but inactive polymerase.

- As in yeast, activation typically results in the appearance of the transcriptional machinery at the promoter. Thus, the transcriptional machinery is in fact recruited by the activator.
- In common with yeast genes, mammalian genes can be expressed in activator bypass experiments. For example, a hybrid protein bearing Gal4's DNA-binding domain fused to TBP creates an activator. For reasons we do not understand, the effect is weak compared to some activator bypass experiments in yeast. As described in the previous chapters (see Activator Bypass Experiments panel in Chapter 2), success of an activator bypass experiment shows that recruitment of the transcriptional machinery suffices for activation.[3]
- Synergistic effects on activation are easily generated in higher eukaryotes. Multiple copies of the same activator work synergistically, as do combinations of different activators—even combinations and arrangements not found naturally. As would be expected from these observations, synergy need not involve cooperative binding of the activators. The simplest interpretation of these results is that the activators contact different surfaces on the transcriptional machinery, with each contact helping recruit the machinery (for a similar experiment in bacteria, see Figure 1.18).

In sum, what evidence we have is consistent with the notion that higher eukaryotic activators work by recruitment.

WHAT IS RECRUITED?

We begin with two observations. First, the transcriptional machinery of higher eukaryotes (not surprisingly) is similar in many respects to that of yeast. Second, as assayed in vitro, activating regions contact many components of that machinery. Here are a few details.

Transcriptional Machinery and Promoters

The 13-subunit polymerase, the general transcription factors (GTFs), and certain TBP-associated factors (TAFs), and nucleosome modifiers found in yeast (see Table 2.1) have homologous counterparts in *Drosophila* and mammals.[4] There are some differences. For example, *Drosophila* and humans encode alternative forms of TBP and certain TAFs. It is believed that these alternative forms are required for transcribing alternative sets of genes, sometimes doing so only in specific cell types.[5]

Some of the components of the mammalian and *Drosophila* mediators have direct counterparts in yeast—Srb7, for example—but many do not. No counterpart of the yeast protein Gal11 is found in mammalian cells, although a possible homolog has been found in *Drosophila* (see Figure 3.2).

Perhaps the best-characterized promoters among all of the eukaryotes are those found in *Drosophila*. These come in two basic classes. Both contain a so-called initiator (Inr) element that includes the transcriptional start site. Some contain a TATA element about 30 bp upstream, whereas others lack a TATA element and instead have a so-called DPE (downstream promoter element) located about 30 bp downstream. Both promoters require TBP.

Nucleosomal Templates

The histone octamer of the nucleosome is virtually identical in all eukaryotes. The additional histone H1 is believed to facilitate higher-order compaction of chromosomes in higher eukaryotes, but its role in yeast is less clear.

The general level of histone acetylation is higher in yeast than in mammalian cells; that is, in mammals, only histones associated with genes being transcribed are as highly acetylated as is the average nucleosome in bulk yeast chromatin (10–12 acetylated lysines per nucleosome). Presumably, this difference reflects a different balance between the background activities of histone acetylases (HATs) and histone deacetylases (HDACs) in the two organisms. (We speculate on the reason for this difference below; see Chromosomal Position and Gene Expression on page 134.) Directed histone acetylation might therefore have a more important role in activating genes of higher organisms than it does in yeast.

Targets

Acidic activators from higher eukaryotes interact with an array of proteins in vitro, as do yeast activators. These potential targets include components

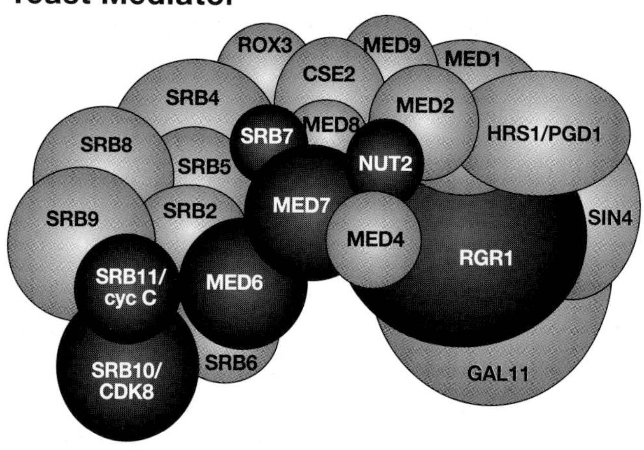

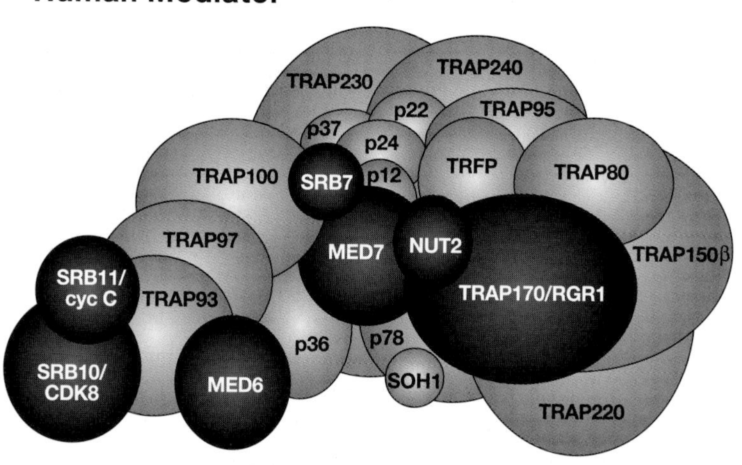

FIGURE 3.2. Yeast and mammalian mediators. Homologous proteins are in black. (Modified, with permission, from Malik and Roeder 2000 [copyright Elsevier Science] [see Chapter 2].)

of mediator, TBP, TAFs, and nucleosome-modifying complexes. Specific targets in the mediator have been proposed for the acidic activators p53 and VP16, for example, and both activators are believed to contact additional components of the machinery as well. We discuss below the possibility that SP1 and other non-acidic activators recognize targets other than those (or a subset of those) recognized by acidic activators (see The *Drosophila HSP70* Gene).[6,7]

Many different activators evidently contact a protein called CBP. This protein, which has a HAT activity, has no counterpart in yeast. CBP is believed to interact with one or more parts of the transcriptional machinery, and some reports classify CBP as part of the mediator. CBP is often referred to as a coactivator, a term applied to many proteins and protein complexes that can be involved in gene activation—nucleosomal modifiers, for example. Because it has so many different meanings in different contexts, we have avoided using this and the analogous term "corepressor."

In Chapter 2, we described experiments that detect the arrival of various components of the transcriptional machinery at a gene upon activation. We now provide some examples of analogous experiments performed with genes of higher eukaryotes.

The Human Interferon-β Gene

This gene is switched on by three activators working together, an example of signal integration discussed below. The activators, as we will see, bind the enhancer cooperatively to form a so-called "enhanceosome." The important point in the current context is that, as analyzed by ChIP experiments, the promoter is vacant prior to activation. Following formation of the enhanceosome, the machinery assembles at the promoter. That machinery includes mediator/polymerase (including CBP); a HAT complex, analogous to yeast SAGA; the human equivalent of the yeast Swi/Snf; and the GTFs, including TBP.[8]

As we noted in Chapter 2, recruitment of multiple complexes presents a mechanistic problem: in this case, for example, does the enhanceosome remain intact, recruiting one component after the other; or does it form, recruit one component, disassemble, and reassemble for the next round of recruitment?

Is has been suggested that the enhanceosome is disassembled by acetylation of proteins within the enhanceosome, a modification mediated by the recruited HATs. This theme has been suggested in other contexts as well. For example, fluorescent microscopy experiments suggest that the activator of the MMTV (mouse mammary tumor virus) promoter—the glucocorticoid receptor—rapidly binds, dissociates, and rebinds as it works. There are also suggestions that some mammalian activators are rapidly ubiquitylated and proteolyzed as they work.

We saw in yeast an example of staged recruitment involving the sequential action of two regulators at the *HO* gene (see page 94). We see another clear example of staged recruitment involving different regulators at the *Drosophila HSP70* gene.

The *Drosophila HSP70* Gene

HSP70 is induced by heat shock; that treatment activates the strong acidic activator HSF (heat shock factor) which then binds to sites upstream of the gene. In the absence of active HSF, a polymerase is found stalled on the gene after having transcribed only 20–40 bp. The presence of that polymerase is believed to depend on the so-called GAGA factor, which binds constitutively upstream of the *HSP70* gene.

Upon heat shock, HSF binds and recruits missing components of the transcription machinery, including the kinase called P-TEFb. P-TEFb phosphorylates polymerase, and (somehow) allows transcription to elongate. There is nothing very special about the HSF activating region: Gal4 activates this gene efficiently when the HSF sites are replaced by Gal4 sites.

The sequence of events at the *HSP70* promoter is mirrored elsewhere. For example, at the promoter of the HIV-1 (human immunodeficiency virus type 1), the activator Sp1, perhaps working with other non-acidic activators, is believed to activate transcription that stalls a short distance downstream. Transcription resumes when a specialized activator called TAT (see Chapter 4) recruits P-TEFb to the promoter. P-TEFb can also be recruited by the acidic activator NFκB.

The generalization seems to be that certain non-acidic activators can recruit only single components of the transcription machinery, whereas acidic activators can recruit them all.[9]

REPRESSION

Like yeast, higher eukaryotes encode specific DNA-binding repressors that counter the effect of activators. Repression can be achieved in two general ways:

- Specific DNA-binding proteins recruit common repressing complexes to different genes. An example is the repressing complex associated with the protein Groucho. Groucho, found in mammals and *Drosophila*, is similar to the yeast repressor Tup1. Groucho, once recruited to a promoter, itself recruits one or more HDACs; and it may also work by the additional mechanisms (not fully understood) that we discussed for Tup1.
- A *Drosophila* protein called CtBP has been suggested to work as a "local" repressor (a "quencher"); i.e., when bound near one group of activators, it prevents their activity without interfering with activators working from different enhancers upstream of the same gene. Although the mechanism is not clear, this form of repression is useful where more than one enhancer controls the same gene in response to different sets of signals.[10]

An additional mechanism of repression, not found in yeast, involves DNA methylation. Methylation of DNA can directly exclude protein binding (at a promoter, for example), and it can create sites that bind proteins that recruit repressing complexes. We encounter examples of DNA methylation later on in this chapter.

DETECTING AND TRANSMITTING PHYSIOLOGICAL SIGNALS

As in yeast, the activities of many higher eukaryotic DNA-binding regulators are controlled by extracellular signals. A wide array of strategies is employed, and discussed below are a few examples.

Transport Into and Out of the Nucleus

Glucocorticoid receptor: This regulator is held in the cytoplasm by a chaperone protein. Interaction with the proper steroid (e.g., dexamethasone) induces a conformational change in the receptor that frees it from the chaperone and allows it to move to the nucleus and bind DNA.

NFκB: This activator is a heterodimer of p65, a protein that bears an acidic activating region, and a second protein called p50. p65 is held in the cytoplasm by the inhibitor protein IκB. Interaction of certain cytokines with their receptors (as well as many other signals) triggers phosphorylation and subsequent degradation of IκB, whereupon p65 moves to the nucleus.[11]

Notch: This activator is a membrane-bound receptor that is cleaved upon interaction with any one of its ligands, which include a protein called Delta. A fragment of Notch, bearing an acidic activating region, moves to the nucleus where it attaches to a DNA-bound protein. This and the next case mediate numerous instances of communication between cells during development.

β-catenin: This activator is ordinarily continuously destroyed by proteolysis in the cytoplasm. When the signal protein Wnt binds to its cell-surface receptor, proteolysis of β-catenin is inhibited. The stabilized protein then moves to the nucleus where it attaches to a DNA-bound protein and activates transcription.

Mef-2: This regulator binds DNA upstream of genes required for myogenesis. In the absence of the signal that triggers myogenesis, an HDAC (HDAC5) binds to Mef-2, and in this configuration, the genes are repressed. Upon receiving the myogenic signal, HDAC5 is phosphorylated, triggering its export from the nucleus, and Mef-2 is left free to activate.

Tubby: This DNA-binding activator is sequestered in the cytoplasm by binding to phosphatidylinositol (PI) lipids in the membrane. Binding of certain hormones activates an enzyme that breaks down the PI lipids, and thereby frees Tubby to move to the nucleus. Mice lacking Tubby, as one might have guessed, are obese.

Phosphorylation of Inhibitor or Activator in the Nucleus

E2F: This DNA-bound protein is inhibited by another protein called Rb (the retinoblastoma protein). Rb covers the E2F activating region and recruits an HDAC to ensure that the genes are switched off. The repressed genes are reactivated only when Rb is phosphorylated, a modification that disrupts its interaction with E2F. This modification, in turn, occurs when cells are ready to enter S phase, and E2F activates genes required for entry into that stage of the cell cycle.

CREB (cAMP receptor element-binding protein): cAMP activates the kinase PKA, which in turn phosphorylates CREB's activating region. This phosphorylation increases the affinity of CREB for CBP (which, please recall, may or may not be part of the mediator).

TRANSPORT-DEPENDENT PROTEOLYSIS: THE CASE OF SREBP

SREBP (sterol-receptor-element-binding protein) binds DNA and turns on genes that encode cholesterol-synthesizing proteins. The activator is held in the membrane of the endoplasmic reticulum (ER) as part of a larger precursor protein (see figure). Another protein (called SCAP) escorts that precursor from the ER to another cellular compartment, the Golgi. The Golgi (but not the ER) contains a protease (called S1P) which cuts the precursor in the loop as shown in the figure. A second protease (S2P) cleaves the precursor once again, releasing SREBP. SREBP moves to the nucleus and activates the appropriate genes.

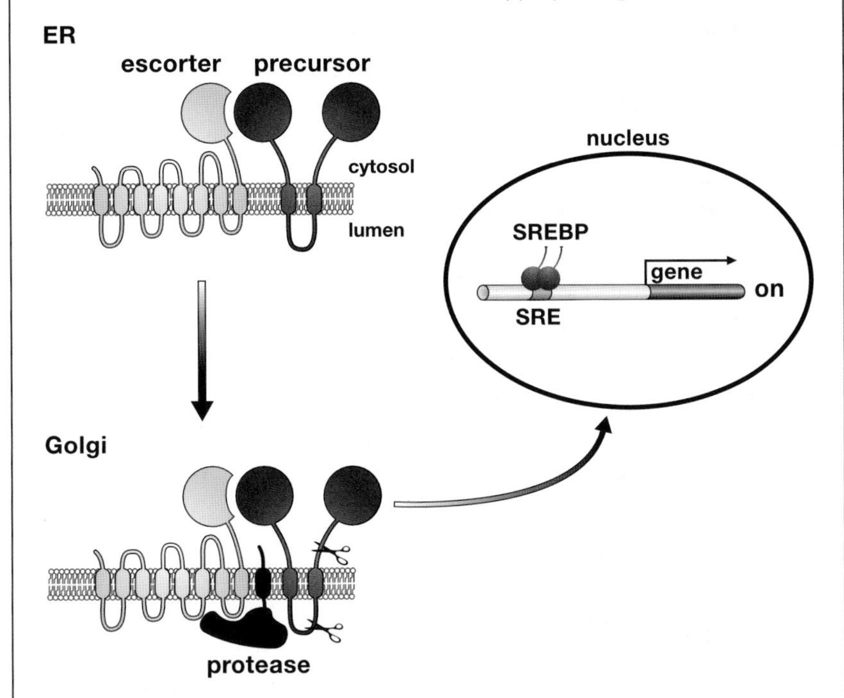

(Modified, with permission, from Brown M.S. and Goldstein J.L. [1999] *Proc. Natl. Acad. Sci. 96:* 11041–11048 [copyright National Academy of Sciences].)

The transporter SCAP is a cholesterol sensor: it moves to the Golgi only if the concentration of cholesterol is low. Thus, the genes encoding the cholesterol-synthesizing enzymes are kept off in the present of cholesterol. Two experiments emphasize that the sole role of SCAP is to relocate the SREBP precursor from one compartment to another.

- Treatment of cells with the chemical brefeldin A switches on SREBP-activated genes. The treatment causes redistribution of proteins, including the protease S1P, from the Golgi to the ER.
- SREBP-activated genes are constitutitively activated by a derivative of S1P that bears a peptide directing it to the ER.

Allosteric Change of DNA-bound Activator

Retinoic acid receptor: This regulator binds its DNA sites constitutively, but in the absence of its ligand—retinoic acid—its activating region is buried. In this state, it recruits an HDAC. Retinoic acid induces a conformational change in the receptor that releases the HDAC and exposes an activating region.

The panel shows an even more elaborate example of regulation of a transcriptional activator.

SIGNAL INTEGRATION, COMBINATORIAL CONTROL, AND ALTERNATIVE ENHANCERS

We introduced this chapter noting that higher eukaryotes exploit signal integration and combinatorial control to a greater degree than yeast. In addition, higher eukaryotes often express the same gene using alternative enhancers in response to alternative signals.

We describe here two cases illustrating these strategies. In the first, the human interferon-β gene, we see how three signals are integrated to switch on that gene. In the second, we see how multiple enhancers control the *Drosophila eve* gene, allowing that gene to be expressed in response to different sets of signals at different times and places during development. Both cases also illustrate combinatorial control, because in each case the activators involved work with other regulators at other genes.[12]

We need invoke no ideas beyond those of cooperative binding and recruitment to explain these cases. It would be difficult to explain many of the following observations were activators to work other than by recruitment.

The Human Interferon-β Enhancer

Viral infection triggers three signal transduction pathways that together switch on the human interferon-β gene. Each pathway transfers one activator (or family of activators) from the cytoplasm to the nucleus. Three activators bind the enhancer to form the enhanceosome (Figure 3.3). The enhanceosome, in turn, recruits the transcriptional machinery (as we have described above) to the interferon-β promoter.[13] Analysis of the human interferon-β enhanceosome is instructive in several regards.

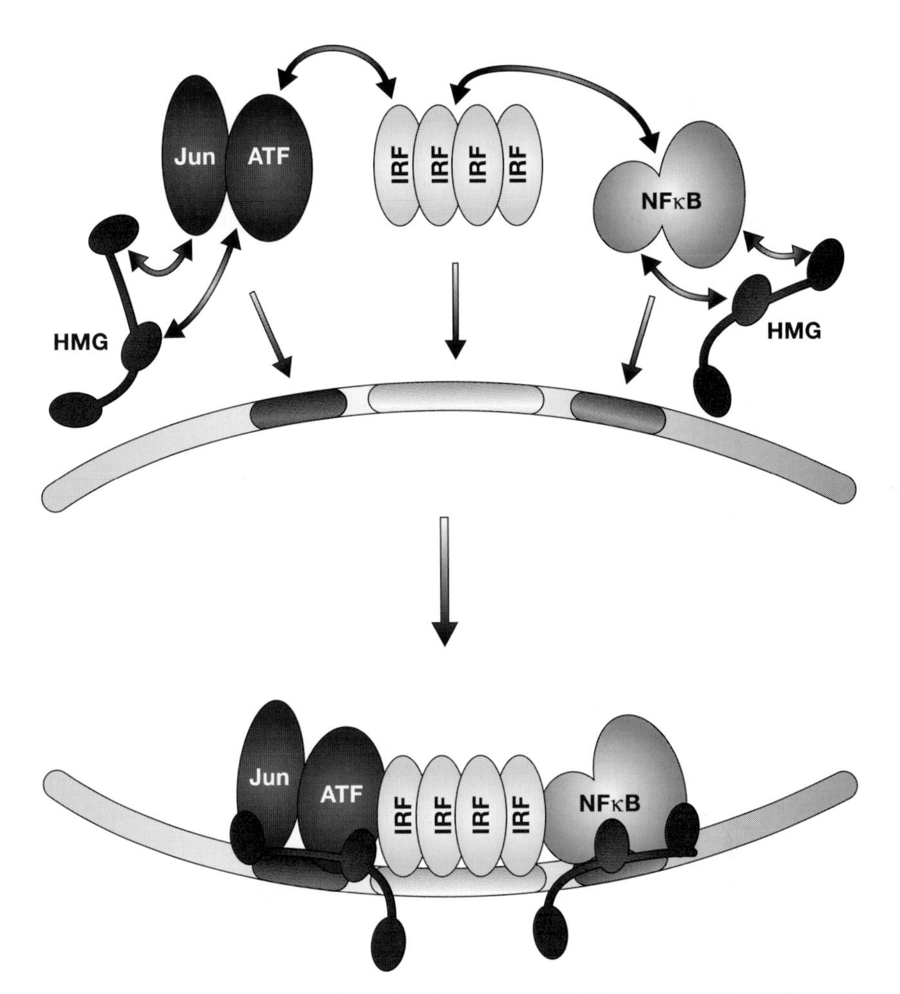

FIGURE 3.3. The human interferon-β enhanceosome. HMG represents HMGI/Y, a ubiquitous protein that binds cooperatively with the three activators. HMGI/Y both bends the DNA and contacts the activators. Each of the transcription factors shown is a member of a family of related activators. In the text, we noted that the enhanceosome might be disassembled and reassembled as it works. It is suspected that the exact composition of the enhanceosome may change as induction proceeds—different members of the IRF family, for example, assembling each time. (Modified, with permission, from Ptashne and Gann 1998 [copyright Elsevier Science] [see Chapter 4].)

- The proteins bind to the enhancer in a highly cooperative manner. Thus, all three signal transduction pathways must be active for the gene to be turned on.

- Each of the three activators also works at other genes in conjunction with other activators, an example of combinatorial control. For example, NFκB is required for activation of many other genes where it works in conjunction with activators different from those found here.
- The enhancer can be replaced by multiple binding sites for just one of the activators. The single activator, present in multiple copies, activates the gene. Thus, the enhanceosome does not present a unique structure required for gene activation.[14]

The *Drosophila eve* Gene

In this section, we consider two of the multiple enhancers associated with the *eve* gene (see Figure 3.4): The stripe 2 enhancer, which directs expression of *eve* in a specific stripe in the early embryo, and the MHE (muscle-heart enhancer) which directs expression, later in development, in cells destined to form part of the heart and the dorsal-most muscles of the fly. Each of these enhancers bears multiple (10–15) binding sites for regulatory proteins. Proteins that bind to the stripe 2 enhancer are different from those that bind the MHE. Thus, no unique set of proteins (nor any single unique protein) is required to activate *eve*.

- The MHE enhancer responds, in a highly restricted region of the late embryo, to the confluence of three signals sent by other cells. Moreover, only some cells have, by virtue of their history, regulatory proteins required for the activity of this enhancer. Thus, specificity is achieved: only certain cells have the potential to respond to the signals (i.e., have

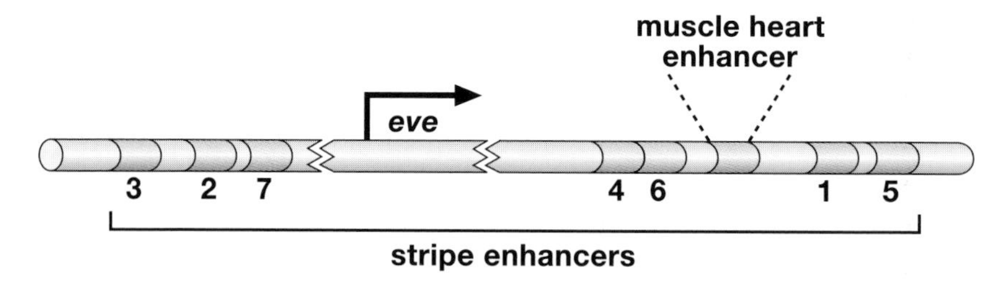

FIGURE 3.4. The *eve* gene. Each of the stripe enhancers activates a stripe of expression of *Eve* in the early embryo (see Figure 3.5). The muscle-heart enhancer is found downstream from the gene, and activates expression in the small area of the late embryo destined to form that organ. Each enhancer spans approximately 150–200 bp and binds 15–20 regulators.

the proper auxilliary activators); of these cells, only those in the right place and right time receive all three signals.[15]

- The stripe 2 enhancer binds proteins—both activators and repressors—that are expressed in overlapping patterns in the early embryo. (At this stage, the embryo is a single large cell with many nuclei, and the regulatory proteins form gradients spanning the embryo.) Only at a certain position in the embryo are the activators (Bicoid and Hunchback) present at high enough concentration, and the repressors (Krüppel and Giant) present at a low enough concentration, that the activators bind and the repressors do not, and the gene is expressed (see Figure 3.5).

We have noted that the several alternative enhancers that control the *eve* gene are dispersed over thousands of base pairs. Recall that in yeast, activators typically work only when positioned within a few hundred base pairs of the gene. How then do these enhancers in higher eukaryotes work "at a distance"?

ACTION AT A DISTANCE

One answer to the above question might be that proteins binding to well-separated sites interact directly, with the intervening DNA looping out to accommodate the interaction. We encountered several such effects in bacteria (see Chapter 1, describing λ O_R-O_L interaction, Lac repression, and NtrC). We also saw how a preformed loop (at a telomere) can facilitate communication between an activator and a distal promoter in yeast.

In an alternative to looping, proteins bound to enhancers might move along the DNA until they reach the promoter. Recall that we encountered such a tracking mechanism in our discussion of phage T4 late gene expression in Chapter 1. In that case, gene activation was linked to DNA replication, a restriction that does not hold for other examples of gene activation that we know of. In addition, the experiments of Figure 3.6 further argue against a tracking mechanism for promoter-enhancer communication in two disparate cases in higher eukaryotes.

- Tracking (in its simplest formulation) requires that the enhancer and promoter be on the same DNA molecule. As shown in Figure 3.6a, an enhancer on one *Drosophila* chromosome can activate a promoter on

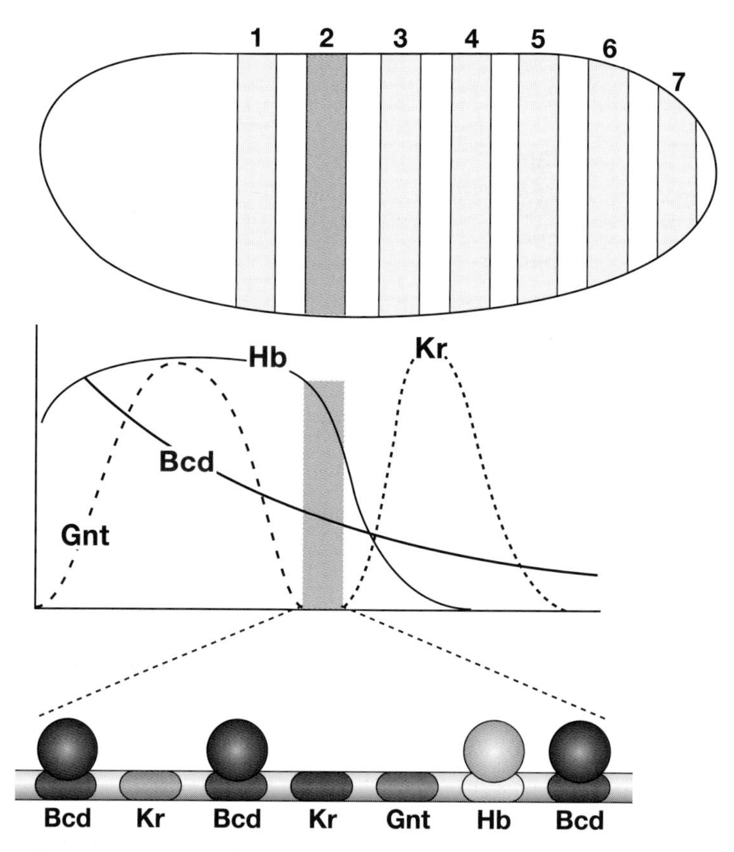

FIGURE 3.5. The *eve* stripe 2 enhancer. A *Drosophila* embryo is shown at the top, with seven stripes where *eve* is expressed. Stripe 2 is provided by the stripe 2 enhancer: it contains binding sites for two activators—Bicoid (Bcd) and Hunchback (Hb)—and two repressors—Krüppel (Kr) and Giant (Gnt), as shown in the bottom part of the figure. The distributions of the regulators across the embryo are shown in the center of the figure. If the Bicoid and Hunchback sites are deleted, the stripe vanishes; if the Krüpple and Giant binding sites are deleted, the stripe widens. (Modified, with permission, from Ptashne 1992 [see Chapter 1].)

another, provided the chromosomes pair. The phenomenon is called transvection.

- Tracking (again in its simplest formulation) would predict that an interruption in the intervening DNA would block promoter-enhancer communication. The experiment in Figure 3.6b shows that an enhancer on one DNA fragment can activate a promoter on another, provided the two DNA fragments are brought together by a protein bridge.

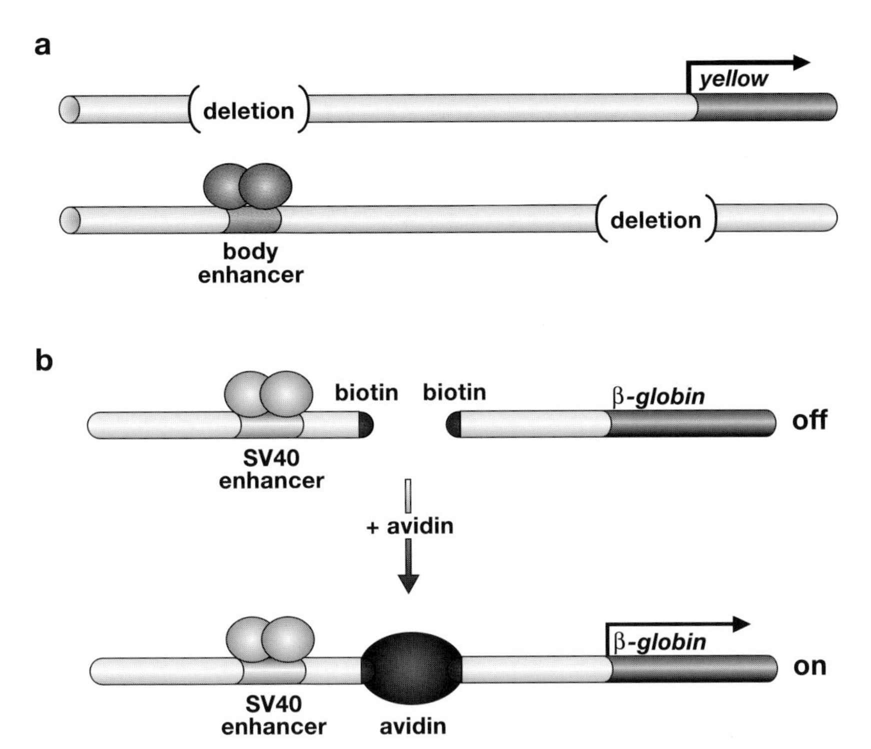

FIGURE 3.6. Two experiments that argue against tracking. (*a*) Transvection. The enhancer of the *yellow* gene on one chromosome can activate the *yellow* gene on a homologous chromosome. In this experiment, the promoter of the *yellow* gene is deleted from one chromosome, and the enhancer is deleted from the other. Activation requires that the chromosomes pair. The result argues against the "tracking" model for enhancer-promoter communication, assuming that a hypothetical tracking activator cannot jump from one chromosome to another. (*a*, Modified, with permission, from Dorsett 1999 [copyright Elsevier Science].) (*b*) Protein Bridge. In this case, the enhancer and promoter on different DNA molecules are brought together by a protein bridge. The bridge is formed by an avidin molecule (a protein) that binds tightly to biotin residues added to the ends of the DNA fragments. The enhancer is from a monkey virus called SV40, and the promoter is from the mammalian β-globin gene; the experiment was performed in vitro. The experiment argues against tracking if it assumed that a hypothetical moving activator cannot hop over the protein bridge. (*b*, Modified, with permission, from Ptashne 1992 [see Chapter 1].)

By arguing against tracking, these experiments return our attention to looping. It is difficult to see how rather weak interactions over very large distances would overcome the entropic barriers to loop formation. Presumably, additional factors facilitate protein interactions over large distances. We have encountered strategies that aid loop formation: in bacteria, the protein IHF can bind DNA between activator and gene and, by bending the DNA, help activation; and, in yeast, as reiterated above, the folded chromatin structure at the telomere can facilitate long-range communication between activator and gene.

Perhaps in higher eukaryotes, the chromatin is more highly folded than in yeast, thereby effectively reducing enhancer-promoter separation. In this regard, recall two observations: nucleosomes in general are less highly acetylated in higher eukaryotes than in yeast; and hypoacetylation is associated with folded structures such as those found at yeast telomeres. Proteins that bind between enhancer and promoter, and themselves interact to form smaller loops, could further reduce the effective distance.[16] The mechanism of long-range enhancer-promoter communication remains an intriguing problem.

DNA METHYLATION, INSULATORS, AND IMPRINTING

Enzymes that methylate DNA are found in mammals, plants, and bacteria, but not in flies, yeast, or worms. The methylases in higher eukaryotes attach methyl groups to cytosine residues in CpG sequences. The added methyl group protrudes into the major groove of the double helix, disrupting binding of some proteins and facilitating binding of others. Methylation of CpGs located in promoter sequences can drastically decrease transcription, presumably by inhibiting the binding of the transcribing machinery.

Several methylases have been identified in mammals, but in no case do we know which is responsible for a given set of methylations, nor do we know how specificity is imposed. Once a DNA sequence is methylated, it is maintained in that state through successive rounds of DNA replication by "maintenance" methylation—hemimethylated DNA is an especially good substrate for further methylation. Methylation has an important role in imprinting, a matter we now discuss, but it also has a more general role, a matter we return to later.

Most mammalian genes are present in two copies, one inherited from the mother and the other inherited from the father. From what we have said, it would be expected that both copies would be expressed identically, because they bear the same control elements and are located at comparable chromosomal positions. For the majority of genes, this is true, but in some cases, one copy of the gene is on and the other copy is off, a pattern perpetuated through cell division. The phenomenon is called "imprinting."

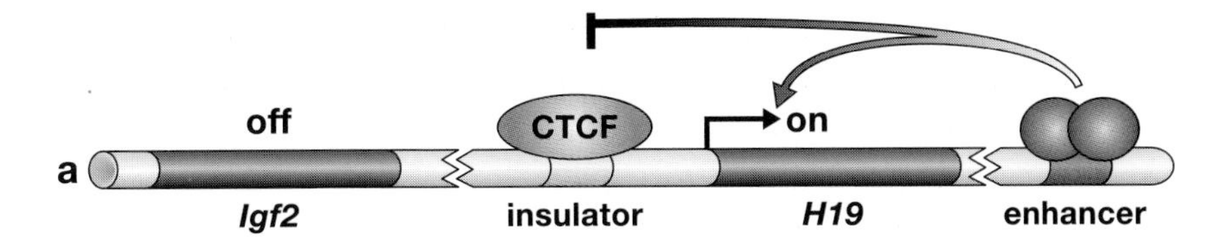

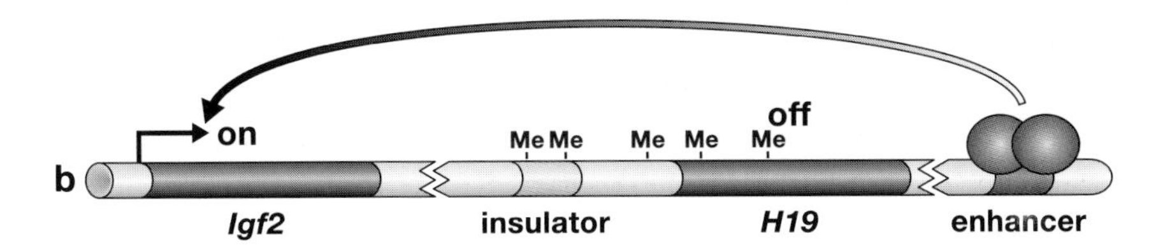

FIGURE 3.7. Imprinting and insulators. The region that includes *Igf2* and *H19* spans about 90 kb. The identity of the protein that binds to the enhancer and activates the genes is not known. "Me" indicates region of DNA methylated. (Modified, with permission, from Wolffe 2000 [copyright Elsevier Science].)

An example of an imprinted gene, expressed only from the maternal chromosome, is *H19* (Figure 3.7). This gene is activated by a downstream enhancer as shown, but only if its promoter is unmethylated. The paternal chromosome is inherited with this promoter region methylated, whereas its maternal counterpart is inherited unmethylated. These states, as we have seen, are readily maintained. Presumably, the umethylated promoter can bind the transcriptional machinery, whereas the methylated promoter cannot.

Figure 3.7 shows a second imprinted gene, *Igf2*, located near *H19*. This gene is activated by the same enhancer that activates *H19*, but it is expressed only from the paternal chromosome. The explanation invokes a protein (called CTCF) that binds a so-called insulator site positioned between the genes as shown. CTCF, like other insulator-binding proteins, blocks communication between an enhancer and a gene—the mechanism by which it does this is not known (but see Footnote 16). CTCF thus prevents the enhancer from activating *Igf2* on the maternal chromosome.

In contrast, on the paternal chromosome, DNA methylation prevents binding of the transcribing machinery to the *H19* promoter and also prevents CTCF binding to the insulator. As a result, *Igf2* is expressed.

CHROMOSOMAL POSITION AND GENE EXPRESSION

We are familiar with the notion that expression of a gene can be affected by its position in the genome. In yeast, a gene ordinarily expressed at a low level can be switched off when positioned near a telomere, and hence in "heterochromatin." This silencing effect can be countered by activators. We noted in passing a similar effect with genes moved near to centromeres—also heterochromatic regions —in *Drosophila*.

An effect of chromosomal positioning on gene expression is frequently encountered when genes are introduced into mammalian cells in culture. In such experiments, genes (with attached enhancers) introduced into cells integrate into the genome at random (so far as we know) to produce stable "transformants." Whether or not the gene is expressed depends on where it has integrated, and even when the gene is initially expressed, this expression can subsequently cease as the cells grow.

When the experiment is performed with a weakened enhancer, fewer of the transformants express the gene, and expression is even less likely to be maintained. Placing an insulator upstream of the enhancer, and another downstream from the gene, has the opposite effect: a higher percentage of integrants initially express the gene, and expression is more stable. Insulators do not substitute for enhancers: genes lacking enhancers are expressed at very low levels if at all, even if the gene is surrounded by the insulators.

So what is the difference between active and inactive genes in these experiments? Comparing an expressed integrant with a nonexpressed integrant reveals that the latter typically bears a higher degree of DNA methylation and less histone acetylation. Both of these conditions are associated with gene repression, and treatment with histone deacetylase inhibitors or methylase inhibitors tends to resuscitate the silenced genes.[17]

Evidently, the enhancer attempts to recruit what is required for transcription, and the methylases/deacetylases tend to make that more difficult. Strengthening the enhancer, or diminishing methylation/deacetylation, tilts the competition toward gene expression. Weakening the enhancer has the opposite effect. All of this—minus the added feature of methylation—is similar to the effects observed with silenced yeast genes.[18]

This, however, raises a further question: why, apparently, can so many locations in the mammalian genome impose silencing, whereas in yeast, only a few regions have this property? Perhaps the explanation lies in the fact that so little of the human genome, compared to that of yeast, encodes useful genes—1% vs. 70%. Much of the excess DNA in mammalian cells comprises "junk" DNA, remnants of viruses and transposons that are best kept quiet. Simply increasing the background activities of HDACs and methylases would have this effect.

Many mammalian genes evidently are protected against silencing by association with regulatory regions called LCRs (locus control regions) in addition to ordinary enhancers. To the extent that they have been analyzed, LCRs seem to comprise a number of regulatory elements, which, in isolation, have properties of insulators, enhancers, and promoters. Thus, we may not need invoke elements other than those we have already encountered to explain the roles of LCRs.[19]

COMPARTMENTALIZATION

In principle, an activator could recruit a gene to the transcriptional machinery without directly interacting with any part of that machinery. Thus, certain compartments of the nucleus could have a particularly high concentration of the machinery, and all an activator would need to do is to bring the gene to that compartment where transcription would then proceed spontaneously. Repressors, in turn, could act by taking the gene to a compartment lacking the transcription machinery or to one bearing an excess of some inhibitory factors.

It is reported that in higher eukaryotes, genes being transcribed tend to be found in parts of the nucleus—near the center—distinct from the locations of repressed genes, which are at least sometimes found near the nuclear periphery. These observations alone do not imply that activators work in any way other than by interacting with the DNA and with the machinery. If indeed the transcriptional machinery is concentrated in particular nuclear compartments, interactions between activator and transcriptional machinery would recruit the gene to those compartments. There is no evidence, to our knowledge, for other kinds of interactions that activate simply by localizing a gene to such a compartment or repress a gene simply by localizing it to a different compartment.[20]

OVERVIEW

Many issues involving gene regulation in higher eukaryotes remain unresolved: the ability of activators to effect recruitment over large distances, for example; how insulators work; the basis of specificity of DNA methylation where it occurs; and various issues we have not discussed, including, for example, the mechanism whereby a single X chromosome in every mammalian human cell, picked at random early in development, is inactivated; and the many examples of "allelic exclusion" wherein a single gene, on one chromosome, is turned on during development, but its allelic partner on the other chromosome is not. Some problems that until recently seemed mysterious, imprinting, for example, are beginning to be understood.

Underlying these answers and problems seems to be a simple mechanistic principle: activating a gene requires bringing the transcriptional machinery to that gene. And it follows that any factors that increase or decrease the ease with which that is accomplished will affect the degree to which the gene is transcribed. The general rule seems to be that the more activators, and the more strongly they recruit the machinery, the greater the degree of gene activation. This is true whether those activators are all the same or are a collection of different activators, working in any combination.

This outlook enables us to make general sense of many gene regulatory phenomena without specifying what the limiting factor in any given case might be. For example, an increase in the background level of DNA methylation or histone deacetylation could render expression of a gene more recalcitrant to the effect of any given set of activators. Overexpression of an activator can lead to its working when it should not: the requirement for cooperative DNA binding with other activators can be obviated, for example, or, at sufficiently high concentrations, the activator can squelch.

The very simplicity of the underlying mechanism explains otherwise puzzling features. For example, there is no obvious "syntax" to regulation. Thus, any given regulator can, evidently, be used to regulate any gene and, conversely, any given gene can be controlled by alternative combinations of regulators.

FOOTNOTES

[1]The glucocorticoid receptor is believed to adopt one conformation when bound to one DNA sequence, and another conformation when bound to a different sequence. In one case, an activating region is exposed, but in the other, it is buried, and instead, the receptor recruits a repressing complex.

[2]Many activators from higher eukaryotes work in yeast, the acidic ones particularly well. Non-acidic activators such as SP1 work weakly on their own but synergize with a weak acidic activator when both are bound in front of a gene. Most families of DNA-binding domains found in yeast are found in higher eukaryotes as well, but the latter also contain additional ones.

[3]Recall that even in yeast, several fusion proteins work (in activator bypass experiments) with greatly differing efficiencies depending on the promoters (see page 80). It may be that an insufficient range of promoter architectures has been tested in the case of higher eukaryotes. A fusion protein bearing one mediator component (human SRB7) was reported to work synergistically in a mammalian cell with a classical activator when both were bound near a gene. The classical activating region had to be tethered to DNA for this effect, consistent with the idea that it was recruiting some additional component(s), rather than working on a prebound complete transcription complex. The finding of a synergistic effect of natural activators with mediator-fusion proteins was not observed in a second study. Whether this apparent contradiction can be explained by differences in the promoters or DNA-binding domains used for the fusions, etc, remains to be seen.

[4]Some of these components are at least partially functionally interchangeable. For example, human TBP will replace yeast TBP for transcription of PolII genes, but not for transcription of certain other genes (transcribed by PolI or PolIII) that also require TBP.

[5]The alternate forms of TBP are called TRF (TBP-related factor) and TLF (TBP-like factor). Yeast have only TBP, and in only one copy.

[6]The proposed target of the thyroid hormone receptor is the mediator protein TRAP220, and that of p53 and VP16 is TRAP80. Consistent with these statements, mouse cells lacking TRAP220 are approximately 80% deficient in thyroid-hormone-mediated activation, whereas VP16 works as well as in wild-type cells.

[7]Initial experiments designed to identify targets of activating regions suggested that different complexes mediate the effects of different activators. Thus, two apparently different protein complexes were isolated that bound the vitamin D and thyroid hormone receptors; these were called, respectively, DRIP and TRAP. Yet another protein complex was found that bound VP16 (called Arc). In other experiments, various multiprotein fractions (e.g., CRSP) were found that facilitate activation in vitro. Finally, a complex (SMCC) was isolated from mammalian cells on the basis of its containing homologs of components of the yeast mediator. Although the issue is not fully resolved, it appears likely that TRAP is identical to SMCC, and other complexes (including DRIP, PC2, ARC, CRISP, and NAT) are subcomplexes of that larg-

er structure which is now referred to as the mammalian mediator. It remains to be seen whether the various smaller complexes are physiologically important or are artifacts of the various isolation procedures used. As for yeast, the large mediator complex can be isolated in two predominant forms, one containing Srb10/11 and the other lacking those proteins. Electron microscopic image analysis suggests that the yeast and mammalian mediators are similar in overall size and shape.

[8]An order of recruitment in vivo has been suggested: SAGA, mediator/polymerase, Swi/Snf, and then TFIID. Activation in vitro, studied with nucleosomal templates, produced the same order of recruitment. The natural interferon-β enhancer can be replaced with an artificial enhancer comprising reiterated sites for just one of the activators. The artificial enhancer was studied in vitro using naked DNA templates. "Order of addition" and other experiments suggested that, in vitro, the assembly of components of the transcriptional machinery proceeds in an order different from that observed in vivo: in this case, TFIID (including TBP) was bound first, then PolII and mediator, and then CBP.

[9]In one study, it was observed that increasing the number of identical acidic activators bound near a gene increased the ratio of elongated to initiated transcripts. This result is consistent with the notion that acidic activators can recruit all of the required components, including those that encourage elongation.

[10]Two explanations have been offered for the effect of a "quencher": First, it could directly inhibit the activating regions of nearby activators, and second, it might "poison" the transcribing machinery when brought to it by those nearby activators.

[11]Some activators in higher eukaryotes, like NFκB, are members of families of proteins that bind as either homodimers or heterodimers in different combinations. Examples include the retinoic acid receptors (which switch genes on and off during development, presumably in response to retinoic acid), the myogenic factors (which control muscle differentiation), and the E2F family (which control entry into the cell cycle). For the most part, so far as is now known, the various alternative forms of each family seem to recognize the same sequence and perform the same function. It would seem likely, however, that there are physiologically differences in their effects that have for most cases not yet been uncovered. Recall that in yeast, α2 can bind as a homodimer to one site or as a heterodimer with **a**1 to another site.

[12]An alternative to the use of multiple enhancers (as at *eve*) would be to generate multiple copies of the gene (by reiterative duplication) and attach single enhancers to each. Both of these approaches are used in higher organisms, but for unknown reasons, the plant *Arabidopsis* seems to have taken predominantly the gene duplication route. This fact helps to explain the surprisingly large number of genes in plants (25,000), nearly the same as that now believed to be found in humans.

[13]Virus infection typically switches on the interferon-β gene in a limited subset of cells. This restriction is caused by formation of inert enhanceosomes that bear, in place of IRF3/7, a protein called IRF2. Cells derived from mutant mice lacking IRF2 respond more homogeneously to viral infection. Thus, IRF2 imposes a kind of negative control on the response to viral infection. Perhaps some cells express a high enough level of IRF2 to be completely nonresponsive to viral infection in this regard.

[14]It is reported that the enhanceosome works somewhat more efficiently than does a limited number of reiterated copies of NFκB.

[15]The signal transduction pathways here are DPP, wingless, and RAS. The three activators that respond to these signal pathways are dTCF, Mad, and pointed, respectively. These activators will work at the MHE only in cells that also bear the activators twist and tinman. Although it is suspected that at least some of the proteins binding to this enhancer do so cooperatively, that cooperativity evidently does not require all of the regulatory protein-binding sites. Thus, for example, deleting certain single sites reduces activation, but does not eliminate activation. Deleting two sites, however, or certain single sites, has a more drastic effect.

[16]Chip, a *Drosophila* protein, has been proposed to form a series of miniloops between enhancer and promoter. An insulator—which disrupts promoter-enhancer communication (see text)—might somehow disrupt formation of these miniloops. It has been reported that pairs of insultors, when suitably positioned, can actually facilitate promoter-enhancer communication, presumably by forming loops that bring those elements closer together.

[17]Both inhibitors are small molecules that readily enter cells. The fact that either treatment works is consistent with the notion that methylation, in this case, is not working by directly blocking access to the gene. Rather, the methylated sequences are probably forming binding sites for a protein (called MePC2) that recruits an HDAC, which in turn effects repression by deacetylating nearby histones.

[18]How the insulators protect against deacetylation/methylation is unclear. Insulators have similar protective effects against heterochromatin-mediated silencing in *Drosophila* where, as we have noted, there is no methylation.

[19]It has been reported that in certain mammalian cells in culture, a gene, even without any known enhancer, can spontaneously turn on to a high level, only to be extinguished as the cells grow. We imagine that this might be caused by a transient change in the concentration of some protein(s), either regulators or the transcriptional machinery, that, perhaps working cooperatively with other proteins, leads to transient activation of the gene. The authors of these experiments suggest that the results should encourage us to think about enhancer action in ways other than those emphasized in this book.

[20]We saw in yeast a case in which artificially moving a partially silenced reporter to the nuclear periphery increased the level of silencing. In this case, the suggested explanation is that the various silenced regions interact and thereby stabilize silencing. No repression was observed, however, if the gene were unable to bind the repressing complex (including Sir2).

BIBLIOGRAPHY

Many of the references given in the chapter on yeast (Chapter 2) are relevant here as well, as are some of the books listed at the end of the Introduction. Listed below are a few additional references.

REVIEWS

Transcriptional Machinery and Promoters

Burke T.W., Willy P.J., Kutach A.K., Butler J.E., and Kadonaga J.T. 1998. The DPE, a conserved downstream core promoter element that is functionally analogous to the TATA box. *Cold Spring Harbor Symp. Quant. Biol.* **63:** 75–82.

Vo N. and Goodman R.H. 2001. CREB-binding protein and p300 in transcriptional regulation. *J. Biol. Chem.* **276:** 13505–13508.

Activation and Repression

Brand A.H., Manoukian A.S., and Perrimon N. 1994. Ectopic expression in *Drosophila. Methods Cell Biol.* **44:** 635–654.

Chen G. and Courey A.J. 2000. Groucho/TLE family proteins and transcriptional repression. *Gene* **249:** 1–16.

Freeman B.C. and Yamamoto K.R. 2001. Continuous recycling: A mechanism for modulatory signal transduction. *Trends Biochem. Sci.* **26:** 285–290.

Lefstin J.A. and Yamamoto K.R. 1998. Allosteric effects of DNA on transcriptional regulators. *Nature* **392:** 885–888.

Lis J. 1998. Promoter-associated pausing in promoter architecture and postinitiation transcriptional regulation. *Cold Spring Harbor Symp. Quant. Biol.* **63:** 347–356.

Parkhurst S.M. 1998. Groucho: Making its Marx as a transcriptional co-repressor. *Trends Genet.* **14:** 130–132.

Enhancers and Action at a Distance

Carey M. 1998. The enhanceosome and transcriptional synergy. *Cell* **92:** 5–8.

Dorsett D. 1999. Distant liaisons: Long-range enhancer-promoter interactions in *Drosophila. Curr. Opin. Genet. Dev.* **9:** 505–514.

Dynan W.S. 1989. Modularity in promoters and enhancers. *Cell* **58:** 1–4.

Jones N.C., Rigby P.W., and Ziff E.B. 1988. Trans-acting protein factors and the regulation of eukaryotic transcription: Lessons from studies on DNA tumor viruses. *Genes Dev.* **2:** 267–281.

Levine M. 1999–2000. Transcriptional control of *Drosophila* embryogenesis. *Harvey Lect.* **95:** 67–83.

Maniatis T., Goodbourn S., and Fischer J.A. 1987. Regulation of inducible and tissue-specific gene expression. *Science* **236:** 1237–1245.

Maniatis T., Falvo J.V., Kim T.H., Kim T.K., Lin C.H., Parekh B.S., and Wathelet M.G. 1998. Structure and function of the interferon-β enhanceosome. *Cold Spring Harbor Symp. Quant. Biol.* **63:** 609–620.

Mannervik M., Nibu Y., Zhang H., and Levine M. 1999. Transcriptional coregulators in developments. *Science* **284:** 606–609.

Merika M. and Thanos D. 2001. Enhanceosomes. *Curr. Opin. Genet. Dev.* **11:** 205–208.

Serfling E.J. and Schaffner W. 1985. Enhancers and eukaryotic gene transcription. *Trends Genet.* **1:** 224–230.

Taniguchi T., Ogasawara K., Takaoka A., and Tanaka N. 2001. IRF family of transcription factors as regulators of host defense. *Annu. Rev. Immunol.* **19:** 623–655.

Physiological Signals

Brown M.S., Ye J., Rawson R.B., and Goldstein J.L. 2000. Regulated intramembrane proteolysis: A control mechanism conserved from bacteria to humans. *Cell* **100:** 391–398.

Cantley L.C. 2001. Translocating tubby. *Science* **292:** 201–2021.

Chuang P.T. and Kornberg T.B. 2000. On the range of hedgehog signaling. *Curr. Opin. Genet. Dev.* **10:** 515–522.

De Cesare D., Fimia G.M., and Sassoni-Corsi P. 1999. Signaling routes to CREM and CREB: Plasticity in transcriptional activation. *Trends Biochem. Sci.* **24:** 281–285.

Dynlacht B.D. 1997. Regulation of transcription by proteins that control the cell cycle. *Nature* **389:** 149–152.

Dyson N. 1998. The regulation of E2F by pRB-family proteins. *Genes Dev.* **12:** 2245–2262.

Glass C.K. and Rosenfeld M.G. 2000. The coregulator exchange in transcriptional functions of nuclear receptors. *Genes Dev.* **14:** 121–141.

Hill C.S. and Treisman R. 1995. Transcriptional regulation by extracellular signals: Mechanism and specificity. *Cell* **80:** 199–211.

Maniatis T. 1999. A ubiquitin ligase complex essential for NFκB, Wnt/Wingless and Hedgehog signaling pathways. *Genes Dev.* **13:** 505–510.

Mayr B. and Montminy M. 2001. Transcriptional regulation by the phosphorylation-dependent factor CREB. *Nat. Rev. Mol. Cell Biol.* **2:** 599–609.

Mercurio F. and Manning A.M. 1999. Multiple signals converge on NFκB. *Curr. Opin. Cell Biol.* **11:** 226–232.

Wodarz A. and Nusse R. 1998. Mechanism of Wnt signaling in development. *Annu. Rev. Cell Dev. Biol.* **14:** 59–88.

DNA Methylation and Imprinting

Bestor T.H. 2000. The DNA methyltransferases of mammals. *Hum. Mol. Genet.* **16:** 2395–2402.

Bird A.P. and Wolffe A.P. 1999. Methylation-induced repression—Belts, braces, and chromatin. *Cell* **99:** 451–454.

Tilghman S.M. 1999. The sins of the fathers and mothers: Genomic imprinting in mammalian development. *Cell* **96:** 185–193.

Wolffe A.P. 2000. Transcriptional control: Imprinting insulation. *Curr. Biol.* **10:** R463–465.

Chromosomal Position

Bell A.C. and Felsenfeld G. 1999. Stopped at the border: Boundaries and insulators. *Curr. Opin. Genet. Dev.* **9:** 191–198.

Bell A.C., West A.G., and Felsenfeld G. 2000. Insulators and boundaries: Versatile regulatory elements in the eukaryotic genome. *Science* **291:** 447–450.

Grosveld F. 1999. Activation by locus control regions? *Curr. Opin. Genet. Dev.* **9:** 152–157.

Higgs D.R. 1998. Do LCRs open chromatin domains? *Cell* **95:** 299–302.

Compartmentalization

Francastel C., Schubeler D., Martin D.I., and Groudine M. 2000. Nuclear compartmentalization and gene activity. *Nat. Rev. Mol. Cell Biol.* **1:** 137–143.

Lemon B. and Tjian R. 2000. Orchestrated response: A symphony of transcription factors for gene control. *Genes Dev.* **14:** 2551–1569.

Other Systems and Other Views on Enhancer Action

Blackwood E.M. and Kadonaga J.T. 1998. Going the distance: A current view of enhancer action. *Science* **281:** 61–63.

Fiering S., Whitelaw E., and Martin D.I. 2000. To be or not to be active: The stochastic nature of enhancer action. *Bioessays* **22:** 381–387.

Pirrotta V. 1998. Polycombing the genome: PcG, trxG, and chromatin silencing. *Cell* **93:** 333–336.

CHAPTER 4

Enzyme Specificity and Regulation

We live, I regret to say, in an age of surfaces.

Oscar Wilde

We began this book by casting the problem of gene regulation as one of controlling enzyme specificity. The active site of *Escherichia coli* RNA polymerase (i.e., that part of the enzyme that synthesizes mRNA) does not determine which gene it will transcribe. Rather, in one widely used strategy (regulated recruitment), binding interactions recruit the enzyme to one or another of its potential substrates (genes). Other enzymes involved in transcription—nucleosome modifiers, for example—are similarly regulated.

This recruitment requires interaction between adhesive (glue-like) surfaces on proteins and DNA. The stereospecific constraints on the positions of these interacting surfaces seem slight; there are many patches on *E. coli* RNA polymerase, for example, that can be used to recruit it to the gene. And often these sites are well separated from the enzyme's active site.

We noted that enzymes such as β-galactosidase do not face this kind of specificity problem. Rather, each such enzyme recognizes a unique substrate, and that recognition is effected solely by residues intimately associated with the active site. Such enzymes, which typically work on small molecules, are regulated (if at all) by allostery.

We first encountered allostery in considering control of Lac repressor by lactose. We noted that a derivative of lactose, binding to the repressor (at a site separate from the DNA-binding surface) induced a conformational change in the protein so that it could no longer bind DNA. We have since used the term allostery to refer to any induced conformational change that affects the activity of a regulatory protein, an enzyme, or even (as we have seen in one case) DNA.[1]

Enzymes that work on large molecules can be subject to both forms of control: recruitement and allostery. In the following cases, we will see examples of constitutively active enzymes that are regulated solely by recruitment (e.g., ubiquitylating enzymes), and others (e.g., many kinases) in which regulated recruitment is accompanied by allosteric regulation. Several of the illuminating experimental approaches employed in the study of gene regulation have been applied to these other cases as well.[2]

UBIQUITYLATION AND PROTEOLYSIS

A protease—present in a complex called the proteosome—degrades many proteins in eukaryotes. Regulation of this process is crucial to cell growth and survival. For example, destruction of certain proteins at specific times is required for progression from one phase of the cell cycle to another. How does the proteosome select the proper substrates to degrade at any given time?

Substrate selection by the proteosome is determined by ubiquitylation. Proteins are modified by attachment of polyubiquitin chains and then are rapidly degraded by the proteosome. Thus, our question becomes: how are the proper proteins selected, under any given set of conditions, for ubiquitylation?

Figure 4.1 illustrates one of several ubiquitylating enzymes found in yeast and other eukaryotes. The enzymatic machinery itself is a multiprotein complex to which substrates are recruited by so-called F-box proteins. There are many F-box proteins that can (individually) attach to the complex. Each F-box protein bears two domains: a common "F-box" domain that binds to the enzyme complex, and a unique domain that binds a specific substrate. F-box proteins are analogous to transcriptional activators that work by recruitment. Both simultaneously bind substrate and enzyme—RNA polymerase and gene in one case, the ubiquitylating complex and target protein in the other.

Figure 4.1a shows an F-box protein in action. In this case, the F-box protein (called Cdc4) binds its target protein (called Sic1) when the latter is phosphorylated, a modification that occurs at a specific time in the cell cycle. (Thus, phosphorylation of one protein creates a docking site for another, a theme we encounter often in the pages ahead.) The F-box protein then brings its target to the ubiquitylating machinery.

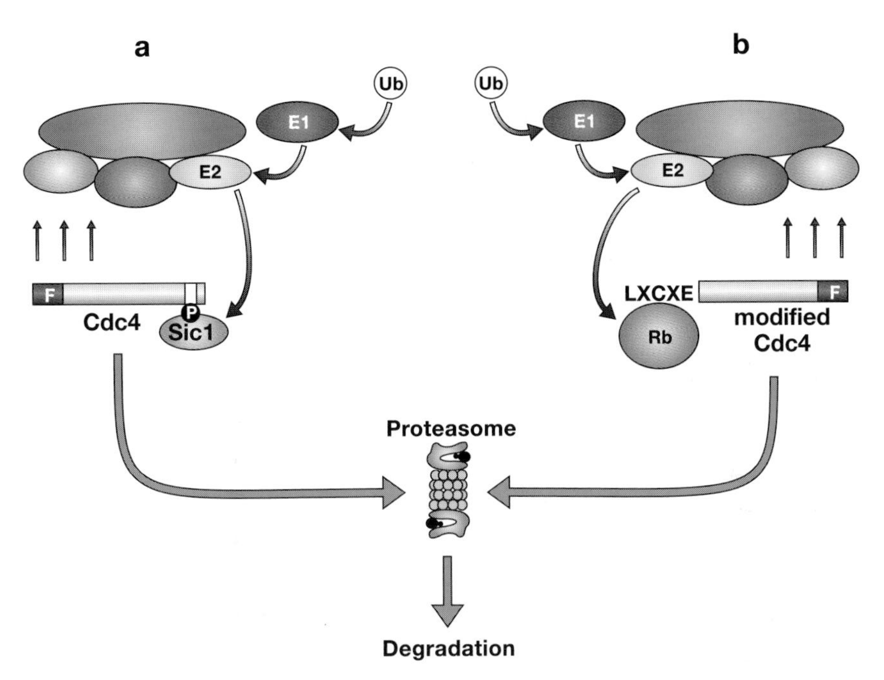

FIGURE 4.1. Recruitment of substrate by F-box proteins to the ubiquitylating machinery. E1 is an enzyme that transfers ubiquitin to E2, and that enzyme in turn transfers ubiquitin chains to substrates recruited by F-box proteins. (*a*) Sic1 is an inhibitor of a cyclin-dependent kinase. When phosphorylated, Sic1 binds the F-box protein Cdc4. Destruction of Sic1 causes progression from one stage of the cell cycle to another. (*b*) The F-box protein from *a* is used here in a modified form: its substrate recognition region has been replaced by a short peptide (of sequence LXCXE) that binds the mammalian protein Rb. Rb is recruited by this hybrid F-box protein to the ubiquitylating machinery, ubiquitylated, and subsequently destroyed. E1 is called a ubiquitylating activating enzyme; E2 (in this case Cdc34) is called a ubiquitin conjugating enzyme, and the remaining four proteins, including the F-box protein, are together called an E3 ubiquitin protein ligase. Ubiquitin is a 76-amino-acid protein and is highly conserved in eukaryotes. (Modified, with permission, from Zhou P. et al. [2000] *Mol. Cell 6:* 751–756 [copyright Elsevier Science].)

Figure 4.1b describes an experiment showing that specificity, i.e., which protein is ubiquitylated, is determined solely by the adhesive interaction of an F-box protein with a target. In this experiment, the substrate-binding portion of the F-box protein (of Figure 4.1a) was substituted by a protein fragment that binds the mammalian protein Rb (retinoblastoma). When expressed in yeast, this mammalian protein is ordinarily stable, but when expressed along with the new F-box protein, it is rapidly ubiquitylated and degraded. Thus, even a protein that is normally not a substrate for the ubiquitylating machinery becomes one when artificially recruited by a modified F-box protein.

The experiment is analogous to various domain swap experiments we have encountered in our study of gene regulation. Swapping one DNA-binding domain for another, for example, preserves the function of a transcriptional activator, but changes the substrate (gene) it controls.[3]

SPLICING

Many genes, especially in higher eukaryotes, bear introns that are transcribed into RNA along with the coding sequences (the exons). These introns must be removed from the RNA to produce a functional mRNA. Just as there is an elaborate machinery for transcribing genes, so is there another machinery for splicing—indeed, some 50 proteins are dedicated to the process–and that process can be regulated.

Characteristic sequences flanking introns are recognized, weakly, by the splicing machinery. Splicing is activated by so-called SR proteins which bind to RNA at nearby "splicing enhancers." Different splicing enhancers bind different SRs, just as different transcriptional enhancers bind different transcriptional activators. The following results show that the SR proteins work (i.e., activate splicing) by recruiting the splicing machinery to a nearby intron (Figure 4.2).

- A given splicing enhancer will work when placed downstream from any intron. This property is analogous to that of transcriptional enhancers, which work on any gene with which they are associated.
- Like typical transcriptional activators, an SR protein bears separable activating and nucleic-acid-binding domains. As for these transcrip-

FIGURE 4.2. Recruitment of the splicing machinery by SR proteins. (*a*) An SR protein is shown binding to a site (a splicing enhancer) in the RNA downstream from the intron to be removed and recruiting the splicing machinery to that intron. The machinery then apposes the 5′ and 3′ splice sites and removes the intron. (*b*) The RNA-binding domain of an SR protein has been replaced by that of MS2 (a bacterial virus protein), and the enhancer has been replaced by an MS2-binding site; recruitment of the splicing machinery then proceeds as in *a*. (*c*) The *Drosophila* SR protein RBP1 binds to the so-called *dsx* splicing enhancer only in the presence of Tra1 and Tra2. Tra1 is expressed only in the presumptive female embryo, and this splicing does not occur in the male-presumptive embryo. (Modified, with permission, from Graveley et al. 1999 [copyright Elsevier Science].)

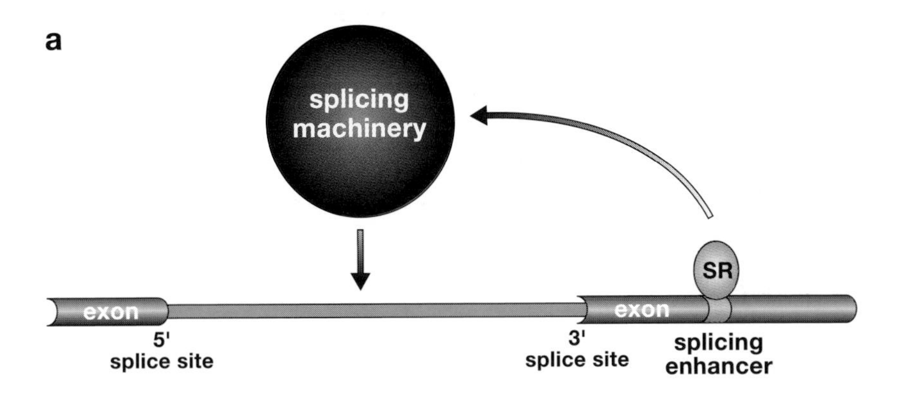

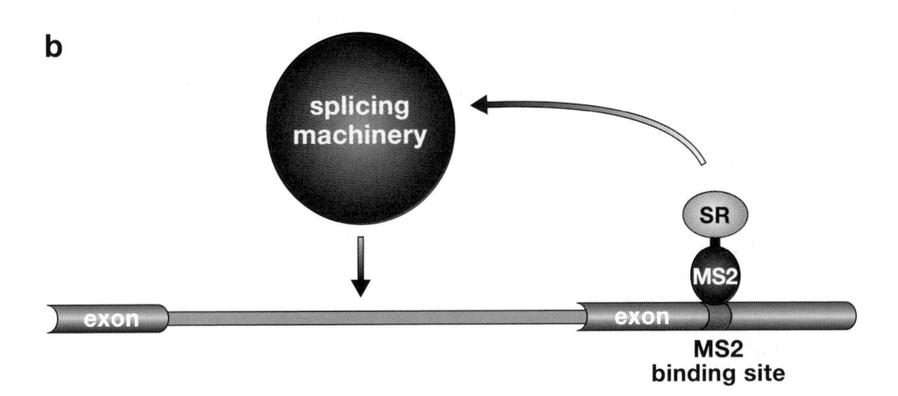

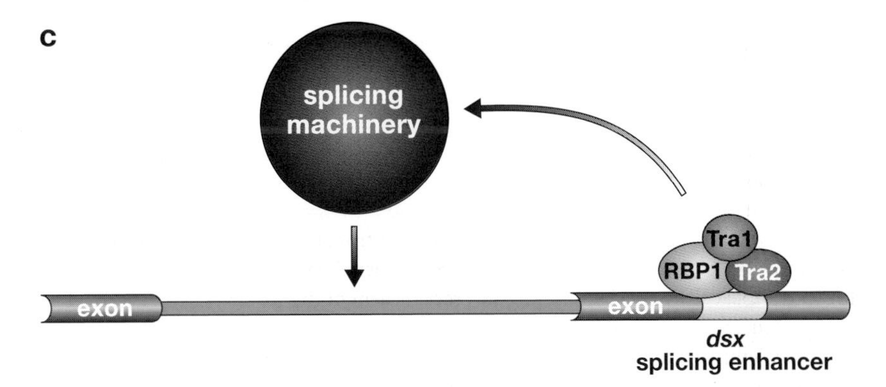

FIGURE 4.2. (*See facing page for legend.*)

tional activators, the role of the nucleic-acid-binding domain (RNA in this case) is simply to position the SR protein near the intron to be spliced, a point illustrated in the domain swap experiment of Figure 4.2b. In that experiment, the RNA-binding domain of an SR protein was replaced by a different RNA-binding domain, and the splicing enhancer was replaced by a sequence recognized by that new domain; splicing was found to proceed as usual.

- The activating regions on the SR proteins resemble their adhesive counterparts on eukaryotic transcriptional activators in that, as assayed using deletion derivatives, they work with an efficiency approximately proportional to their lengths.
- SR-activating regions expressed without RNA-binding domains do not activate splicing.

Many SRs are present ubiquitously. But some, like many transcriptional regulators, are active (or present) only under specified conditions, thus ensuring that the same RNA is spliced differently under different circumstances. For example, the early *Drosophila* embryo develops into a male or a female depending on whether or not a specific intron is removed from a particular RNA molecule. This splicing requires the Tra1 protein, which helps a particular SR bind to the nearby splicing enhancer (Figure 4.2c). Presumptive females, but not males, express the Tra1 protein.

IMPOSING SPECIFICITY ON KINASES

Protein kinases have crucial roles in many aspects of cell growth, and in particular, in conveying the effects of extracellular signals. The most common of these kinases are distinguished by whether they add a phosphate to tyrosine residues or to serine/threonine residues, although some modify both kinds of residues.[4]

The specificity of action of kinases—which tyrosine residues are phosphorylated by a given tyrosine kinase, for example—is determined by two factors.

- The catalytic site of the enzyme prefers a tyrosine found in one sequence context rather than in another. Such contexts include only a few residues around the site phosphorylated, however. This level of specificity may allow a given kinase to distinguish one or another specific target residue out of those available on any given protein. But that specificity alone is not sufficient to pick out the relevant residue(s), on the proper protein, from among the vast array of potential target residues found on all the proteins in the cell.

- Other surfaces on the enzyme, not closely associated with the active site, direct it to the proper target protein. In some cases, this interaction is direct, and in other cases, it is facilitated by an intermediary recruiting protein.

Many kinases are also controlled allosterically: their activities are switched on or off by binding other proteins. Also, many, perhaps all, must themselves be phosphorylated for full activity. A so-called "activating loop" is found near the active site in many kinases. Phosphorylation of that loop is required for full enzymatic activity. Here are some examples in which recruitment determines specificity of kinases.

Cyclin-dependent Kinases

Cyclin-dependent kinases (Cdks) work only when complexed with one or another so-called cyclin protein. The yeast kinase Cdc28, for example, associates with different cyclins at different stages of the cell cycle. Each cyclin induces a conformational change (the same in each case) in Cdc28 that activates the kinase function. In addition, it is believed that each cyclin recruits specific substrates, different for each cyclin, to the enzyme. That change in specificity triggers transitions from one stage of the cell cycle to another.[5]

An example of substrate selection determined by a cyclin is shown in Figure 4.3a. Human cyclin A binds the kinase Cdc28 and directs it to phosphorylate the protein p107. This substrate selection is effected by a hydrophobic patch (HP) on the cyclin as shown. Mutation of that HP reduces phosphorylation of p107 (Figure 4.3b).

The experiment in Figure 4.3c shows that this kinase can be brought to the substrate by a heterologous protein-protein interaction without loss of activity. In this case, the HP patch on the cyclin is destroyed, and a peptide that interacts with the substrate in a novel way is added to the cyclin. The experiment is analogous to various of the domain swap experiments we encountered in our study of gene regulation.

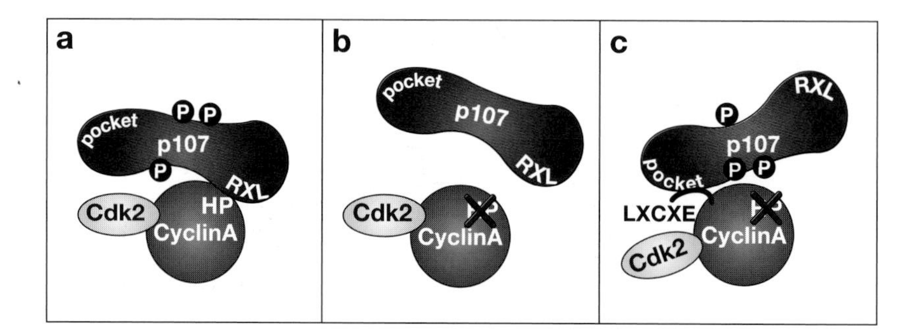

FIGURE 4.3 Recruitment of substrate of a CDK by a cyclin. (*a*) Human cyclin A binds and activates (by changing its conformation) the kinase Cdk2 and, simultaneously, recruits the substrate p107. HP indicates the hydrophobic patch on the cyclin that binds the sequence RXL on p107. (*b*) Removal of the "recruiting" patch on the cyclin abolishes the reaction under the conditions of the experiment (performed in vitro). (c) A new recruiting peptide fused to the cyclin recruits p107 and the reaction is restored. The new recruiting patch used here, LXCXE, is the same as that used in the experiment of Figure 4.1; p107 is a close relative of Rb. Note that the site of interaction with p107 is quite different in *a* and *c*, but the same pattern of phosphorylation is nevertheless observed. (Modified, with permission, from Schulman B.A. et al. [1998] *Proc. Natl. Acad. Sci. 95:* 10453–10458 [copyright National Academy of Sciences].)

Transcriptional Antitermination by TAT

The RNA-binding protein TAT, encoded by the human immunodeficiency virus type 1 (HIV-1), works as a transcriptional antiterminator as mentioned in Chapter 3. In the absence of TAT, polymerase stalls soon after initiating transcription from the HIV promoter. TAT binds a sequence, called TAR, within the resulting short RNA, and recruits a Cdk/cyclin pair to the stalled polymerase. Phosphorylation of the polymerase then allows it to continue transcribing.

The recruited Cdk/cyclin is called P-TEFb, and its components are the kinase Cdk9 and the cyclin CycT1. The phosphorylation is on serine and threonine residues in the carboxy-terminal tail (CTD) of the largest polymerase subunit. Other inhibitory proteins associated with the polymerase are believed to be phosphorylated as well.

The TAT-bypass experiment of Figure 4.4 indicates that the sole role of TAT is to recruit P-TEFb to the stalled polymerase. As shown in Figure 4.4a, TAT has two binding surfaces: one contacts RNA (at the TAR element) and the other contacts P-TEFb. But as shown in Figure 4.4b, TAT

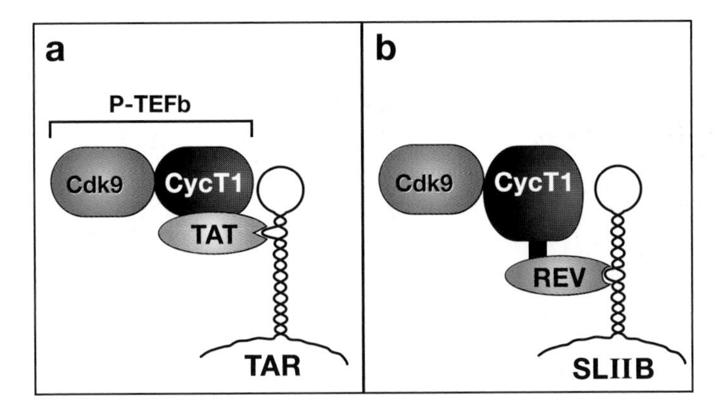

FIGURE 4.4. Recruitment of a Cdk/cyclin pair by TAT. (*a*) TAT, bound to the TAR sequence in RNA, is shown recruiting the cyclin kinase pair P-TEFb by contacting the cyclin component (CycT1) which itself is bound to the kinase Cdk9. The result (not shown) is phosphorylation by Cdk9 of the carboxy-terminal domain (CTD) of polymerase and "antitermination" of transcription. P-TEFb may also phosphorylate other substrates to help elongation. (*b*) A TAT bypass experiment. Rev, another HIV-encoded RNA-binding protein, is fused directly to the cyclin, and the TAR site has been replaced with a REV-binding site (SLIIB). Antitermination is again observed.

can be dispensed with by fusing a different RNA-binding domain to P-TEFb and providing the corresponding RNA-binding site in place of TAR.

Tethering of P-TEFb to DNA at the promoter also works. For example, antitermination is observed when CycT1 (the cyclin component of P-TEFb) is fused to the DNA-binding domain of Gal4, and a Gal4 site is added to the HIV promoter.

The latter experiment is revealing in another way as we now describe. Cdk9 (the kinase in P-TEFb) can be found associated with cyclins other than CycT1, including cyclin K. Fusion of any of these other cyclins to Gal4 does not, however, promote antitermination from a promoter bearing the Gal4 sites. The explanation suggested is that CycT1, unlike the other cyclins, binds the CTD substrate on the polymerase, a surmise supported by the following experiment. The patch on CycT1 believed to bind the CTD was transferred to cyclin K, and that hybrid protein, when fused to the DNA-binding domain of Gal4, promoted antitermination.

We thus see two ways that recruitment determines specificity in this example. The cyclin (CycT1) selects the CTD of polymerase over other potential substrates. But there are many polymerases in the cell, including many bound at different genes. TAT, by recruiting P-TEFb to a particular gene (i.e., one whose transcript includes a TAR sequence), makes the choice among those potential substrates.

Cytokine Receptors and Signaling through STATs

Cytokines are secreted proteins that bind to receptors on cell surfaces, leading to the expression of specific genes. In many cases, the products of those genes are required for an inflammatory response. For example, human interferon is a cytokine produced by cells in response to viral infection. It binds to receptors on the surface of cells surrounding the infected cell and induces the expression of genes whose products inhibit further viral growth.

Cytokines activate transcriptional regulators called STATs. Binding of one cytokine to its receptor activates one STAT, and binding of another cytokine to its receptor activates a different STAT. The various activated STATs bind to DNA and activate different sets of genes.

Signaling proceeds, in outline, as follows (see Figure 4.5). The cytokine binds to the extracellular domain of two receptor chains, bringing those chains together. Each chain has, bound to its intracellular domain, a so-called JAK kinase, and when brought together, these kinases activate each other by cross-phosphorylation. The activated JAKs then phosphorylate a site—a specific tyrosine residue—on the intracellular domain of the receptor.

Phosphorylation of the receptor creates a docking site for a specific STAT (STAT A in Figure 4.5). The bound STAT is then itself phosphorylated by the JAK kinase. Once again, the phosphorylation creates a docking site; in this case, a site that promotes STAT dimerization. When the STATs bind each other, they evidently undergo a conformational change that reveals a nuclear localization sequence. The STAT dimers move to the nucleus, bind specific sites on DNA, and activate specific genes. Other cytokine receptors work similarly, but activate different STATs that regulate different genes.

FIGURE 4.5. A simplified view of cytokine signaling. The steps leading to activation of a specific STAT are shown, followed by gene activation by the activated STAT. One key point is that the specificity of any given JAK—which STAT it phosphorylates—is determined solely by which STAT is recruited to the receptor.

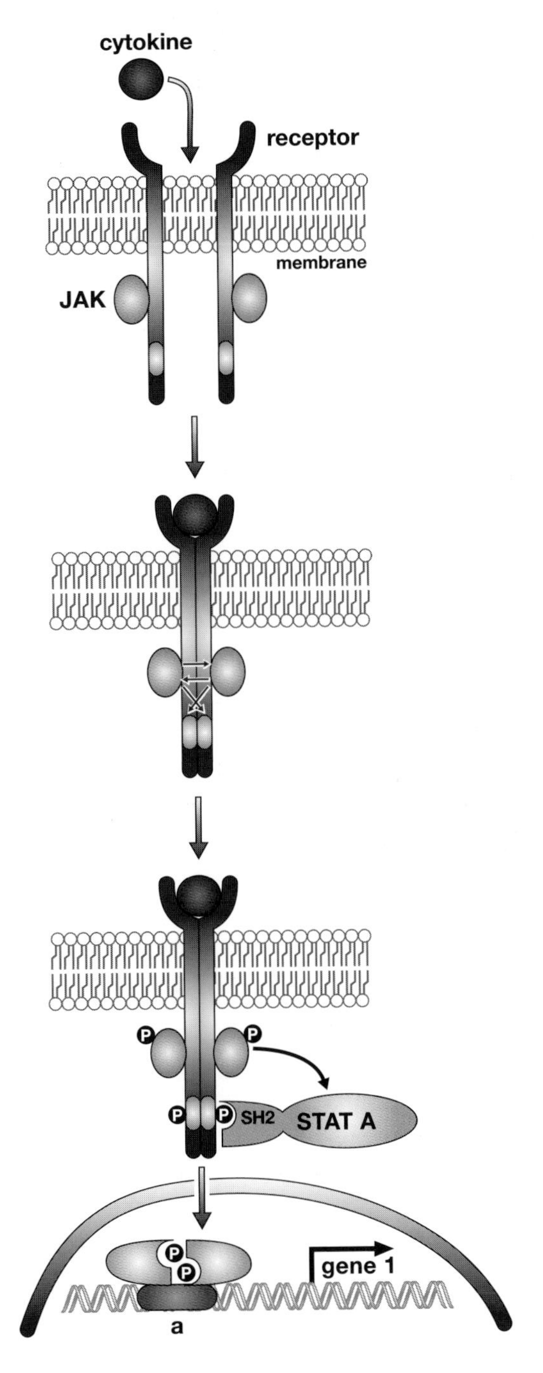

FIGURE 4.5. (*See facing page for legend.*)

Specificity of cytokine signaling—which STAT is activated—is determined by recruitment. That is, the activated JAK phosphorylates whichever STAT is recruited to the receptor bearing that activated JAK. This conclusion is based on various experiments including the following.

- The region on STAT A (in Figure 4.5) that binds the intracellular domain of receptor A can be replaced with the corresponding part of another STAT, STAT B. The hybrid protein is recruited to, and activated by, receptor B.
- Simply overproducing a JAK with a STAT, even one not normally phosphorylated by that JAK, results in STAT activation.

The sites on the STATs that recognize the tyrosine-phosphorylated patches on the receptors are called SH2 domains. These SH2 domains are functionally analogous to the DNA-binding domains of transcriptional regulators. Just as different helix-turn-helix (HTH) domains recognize different DNA sequences, for example, so too do different SH2 domains recognize different target sequences in proteins. These domains are used to bring various pairs of proteins together in a phosphorylation-dependent fashion. The interactions are "glue-like" but differ in specificity—each SH2 domain prefers a phosphorylated tyrosine residue in a specific sequence context of about four amino acids.[6]

Growth Factor Receptors

Proteins called growth factors stimulate the division of cells during development and wound healing. Each growth factor, like each cytokine, binds to an extracellular part of a receptor. The cytoplasmic domains of growth factor receptors are themselves tyrosine kinases. The path from receptor to gene is more complicated than in the cytokine case, as we now describe.

Consider the epidermal growth factor (EGF) receptor. Binding of EGF to its receptor initiates a chain reaction that results in activation of a MAP kinase that, in turn, phosphorylates one or more transcription factors. The key events in transmitting the signal are as follows (see Figure 4.6).

EGF activates the pathway by bringing together two receptor chains. That this is the sole role of the ligand is shown by two experiments: either overexpressing the receptor chains in the absence of EGF, or tethering the chains together artificially, suffices to activate the pathway. As with the cytokine receptors, the apposed chains phosphorylate each other.[7]

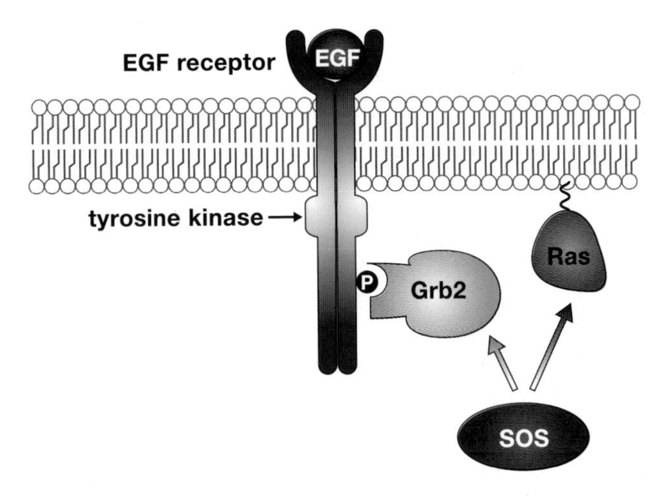

FIGURE 4.6. A simplified view of a generic growth factor receptor. The growth factor EGF brings together two receptor chains, each of which bears an intrinsic tyrosine kinase. Apposition results in kinase activation and phosphorylation of a tyrosine residue on the receptor chain. The modified receptor recruits the adaptor protein Grb2, which in turn binds SOS, and thereby triggers the pathway.

Phosphorylation creates a docking site for an SH2 domain on the "adaptor" protein Grb2. This protein, in turn, brings SOS to the receptor and thereby near the membrane. This relocalization of SOS juxtaposes it with Ras, the next player in the chain, which itself is always tethered to the membrane.

That the sole role of Grb2 is to recruit SOS to the membrane is shown by the following three experiments. These experiments use a truncated form of SOS called SOS′—we will explain why in a moment.

- Overexpressing SOS′ activates the pathway even in the absence of the ligand usually required to activate that pathway.
- Artificially tethering SOS′ to the cell membrane also activates the pathway even in the absence of the ligand.
- Fusion of the SH2 domain of Grb2 to SOS′ obviates the need for Grb2—a Grb2 bypass experiment.

Why was a truncated version of SOS used in the various bypass experiments described above? The answer illustrates a strategy that is widely used to help prevent proteins like SOS from working spontaneously—that is, from working in the absence of the physiologically relevant signal. SOS contains an autoinhibitory flap that obstructs its interaction with Ras. Grb2 interacts with SOS in such a fashion as to displace the inhibitory flap and, as we have emphasized, recruit it to the membrane. We discuss this general theme further later in this chapter. How Ras is activated, and in turn activates the MAP kinases, is discussed in Footnote 8.

Many receptors are more complicated than those we have described. But the basis of this increased complexity is the accumulation of ever more recruiting patches (docking sites) for kinases, substrates, and adaptors. For example, although we have treated the cytokine and growth factor receptors as triggering separate pathways (STAT and Ras, respectively), this is not always the case. There are cytokine and growth factor receptors that can stimulate both pathways. In these cases, a receptor bears the appropriate tyrosine residues which, when phosphorylated, recruit a STAT and (using an adaptor) SOS.

INTERIM SUMMARY AND EXTENSIONS

In this chapter, we have encountered a number of regulatory proteins that resemble transcriptional activators that work by recruitment: they bring together an enzyme with one or another substrate, thereby helping impose specificity. For example, F-box proteins bring target proteins to the ubiquitylating machinery, SR proteins bring the RNA splicing machinery to introns, Grb2 brings a component of a signal transduction pathway to the membrane, and TAT brings a kinase to RNA polymerase.

In these examples, the recruiting protein contains one domain that interacts with substrate, and another that interacts with the enzyme. Both domains serve a "glue-like" function, and in this regard are analogous to the DNA-binding and activating regions of transcriptional activators that work by recruitment. We have encountered a case where a recruiting protein also allosterically activates the enzyme—the effect of a cyclin on its Cdk—a matter we return to below.

Just as there are families of DNA-binding domains, so there are families of protein-protein-binding domains—SH2 domains, for example. Other such domains include PTBs, which, like SH2s, recognize sequences containing phosphorylated tyrosine residues; FHA domains, which see phosphorylated threonine residues; and SH3 domains, whose target sequences (which are not phosphorylated) contain a characteristic array of proline residues.

Docking sites for protein recognition domains are often created by phosphorylation; the sites recognized by SH2 domains are an example. Phosphorylation can also be used to create docking sites for membrane-binding domains. For example, The PI3 kinase (itself recruited to the membrane by binding of its SH2 domain to a phosphotyrosine on an activated receptor) produces the phosphoinositide PIP3 in the region of the membrane nearby. Certain kinases bearing the so-called PH domain bind PIP3, setting off another signal transduction pathway.

The various protein and membrane recognition domains are short—not more than 100 amino acids—and they work when inserted at any of a variety of positions on the surfaces of proteins. Likewise, the typical DNA-binding domain works when inserted at any of a variety of positions within a protein sequence. The DNA-binding domain of Gal4, for example, is normally found at the amino-terminal end of that protein, but it also works when fused at the carboxyl terminus, or when inserted at any of a variety of places within that or other proteins.

Often (it is believed) two or more recruiting domains are used together (cooperatively) to increase affinity and specificity just as, often, transcriptional regulatory proteins bind cooperatively to DNA. For example, in the case of Figure 4.6, although not shown, SOS has a proline-rich sequence recognized by an SH3 domain on Grb2 (thereby linking it to the receptor), and it also has a PH domain that simultaneously interacts with a PIP3 in the nearby membrane.

Kinases

An active kinase is a potentially dangerous enzyme because its active site often can work promiscuously. And so kinases tend to be activated (allosterically) only when in the presence of substrate they are meant to modify. For example, kinases associated with receptors activate themselves by cross-phosphorylation when the receptor chains are brought together. And Cdks are also simultaneously activated and directed to substrate by cyclins. The activities of members of the Src family of kinases are even more closely linked to substrate recognition as follows.

Src is held in an inactive conformation by an intramolecular interaction. As shown in Figure 4.7, SH2 and SH3 domains fold back on the

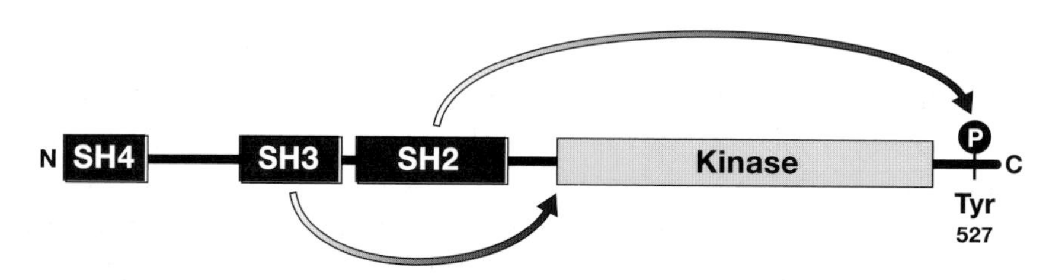

FIGURE 4.7. The Src kinase. A schematic depiction of the kinase (actually Lck, a close relative of Src) in its inactive state. The internal SH2 domain binds to the phosphorylated tyrosine and, cooperatively, the SH3 domain interacts with a proline-rich region as indicated. The SH4 domain includes a site that is modified by myristoylation; this modification anchors the protein to the cell membrane. (Modified, with permission, from Nguyen J.T. and Lim W.A. [1997] *Nat. Struct. Biol. 4:* 256–260 [copyright Nature Publishing Group].)

domain bearing the kinase active site. The SH2 domain interacts with a phosphorylated tyrosine, and the SH3 domain interacts with a proline-containing region. Disrupting these interactions relieves inhibition. One way this is effected is by competition: a substrate that bears higher-affinity sites for the SH2 and/or SH3 domains binds these and unfolds the protein.[9]

Activation of Src at the PDGF (platelet-derived growth factor) receptor illustrates some of these mechanisms. A phosphorylated tyrosine on that receptor binds Src's SH2 domain with high affinity and thereby opens Src. Phosphorylation of Src's activating loop then follows, probably mediated by two juxtaposed Srcs, one on each receptor chain.

Many other ways to control kinase activities are known. For example, the kinase PKA is held in an inactive form by inhibitory subunits. These inhibitory subunits also interact with proteins called AKAPs, which tether the complex to one or another cellular location near potential substrates. In the presence of cAMP, the inhibitory subunits are released.

Phosphatases

These enzymes can also be regulated by recruitment, sometimes working on alternative substrates. For example, the phosphatase called PEST selects one substrate (called p130cas) by recognizing an SH3 domain in that substrate. But an auxiliary recruiting protein called PIP1 can appose the enzyme with a different substrate, the Abl kinase.

Phosphatases have at least three important roles in signaling.

- Phosphatases, recruited to receptors, down-regulate the activating effects of phosphorylation. For example, a phosphatase recruited to the erythropoietin (EPO) receptor limits red blood cell development stimulated by phosphorylation of that receptor. An interesting illustration of this effect was revealed by analysis of a Finnish athlete whose red-blood cell level was aberrantly high (a useful trait when working at high altitudes). He was found to have lost, by mutation, the phosphatase recruiting site on his Epo receptor.[10]

  Other phosphatases remove phosphates from lipids that form membrane-docking sites for proteins. For example, PTEN destroys PIP3 by removing a phosphate from it. In the absence of PTEN, PIP3 accumulates, and the resultant hyper-signaling causes malignancy. PTEN is thus called a "tumor suppressor."
- When recruited to certain receptors, phosphatases can trigger kinase activity. Such an effect is seen in T-cell receptor signaling where a Src-like kinase (Lck) is activated by a phosphatase. The phosphatase removes the phosphate that mediates the inhibitory intramolecular folding of the kinase illustrated in Figure 4.7.
- Certain phosphatases work as part of a background activity to inhibit "basal" levels of activity of various receptors.

The background activities of phosphatases and kinases perform functions analogous to those of the histone acetylases (HATs) and histone deacetylases (HDAC) that work as part of the background that influences the level of expression of genes. Treatment of cells with vanadate, a phosphatase inhibitor, triggers certain signal transduction pathways, just as treatment of cells with HDAC inhibitors increases expression of certain genes in eukaryotes.

Interpreting Signals

We mentioned in the Introduction that the same signal can switch on (or off) different genes at different stages of development. There are several ways this might be achieved, two of which are as follows.

- As mentioned above, activation of a single cytokine receptor can activate multiple signal transduction pathways, a STAT and a Ras pathway, for example. If the STAT (or the Ras component) is present in one cell but not another, the cytokine would have quite different meanings in the two contexts. Another possibility is that the transcription factors that are available to a given MAP kinase (at the end of the Ras pathway) could vary from cell to cell. To what extent these variations occur is under investigation.

- Signals can be integrated on the DNA, and this may in fact be the predominant mode of integration. Thus, a given regulatory protein, activated by a signal transduction pathway, will have different effects depending on the other transcriptional regulators active in the cell. We have seen examples of this throughout our survey of gene regulation—at the *lac* genes in bacteria, the interferon-β gene in humans, and the *eve* gene in flies, to take three examples. Another salient example is provided by the SMAD system. This system mediates signaling by the protein-serine kinase activity of the TGF-β receptor and works in outline very like the STAT system. The same SMAD (a transcriptional activator) activates one set of genes in one cell, but another in another cell, by virtue of its interaction with alternative DNA-binding proteins.

FURTHER GENERALIZATIONS

Many enzymes beyond those we have discussed are regulated by recruitment. These include telomerase (the enzyme that synthesizes the sequence found at the ends of yeast chromosomes); other DNA polymerases, which are brought to proper DNA sites for the initiation of replication and for repair of damaged DNA; caspases, the proteases that trigger apoptosis in eukaryotes; the machinery that translates mRNA into protein, which is recruited to mRNA by a group of proteins that bind cooperatively to sites found both near the beginning and end of the message; and so on.

Here are some consequences of using recruitment to regulate enzyme specificity.

Dangers

Proper regulation can easily go awry. Recruitment determines enzyme specificity by, in effect, increasing the concentration of enzyme near a potential substrate. Thus, if the concentration of an active form of the enzyme were to increase, specificity would tend to be lost. Enzyme concentrations are therefore often maintained at low levels (e.g., *E. coli* RNA polymerase); enzymes are switched on only when their activities are needed (e.g., many kinases); and inhibitors or repressors block access to specific substrates (e.g., Lac repressor).[11]

The concentrations of regulators are often similarly controlled. A transcriptional activator that works by recruitment, for example, works properly only if maintained within certain ranges of concentration. If present at high concentration, it could activate genes inappropriately, having lost its dependence on interaction with other proteins to bind to DNA sites. At sufficiently high concentrations, it would squelch. And so transcriptional activators often negatively regulate transcription of their own genes (e.g., λ repressor); and F-box proteins, whose activities we have likened to those of certain transcriptional activators, are continually degraded, thereby avoiding squelching of the ubiquitylating enzymes that otherwise would likely occur.

Compartmentalization is used both to promote specific reactions and to prevent unwanted ones. Eukaryotic transcriptional activators, for example, are often excluded from the nucleus when they are not needed, and, to take another example, specificity is imposed upon members of the MAP kinase cascade by confining them to scaffolds.

When these controls go awry, the consequences can be disastrous. Described below are four cases of human disease caused by inappropriate activation of a transcriptional activator.

- Human mesothelioma is characterized by an overexpression of the PDGF receptor, which spontaneously dimerizes and activates its associated kinases. This leads to recruitment and hyperactivation of Src, which, in turn, produces an unusually high level of activated STAT3, another protein that is recruited to the PDGF receptor.[12]

- A common form of colon cancer is caused by mutations that interfere with the proteolysis of the transcriptional activator β-catenin. β-catenin, which bears an acidic activating region, attaches to a DNA-bound protein TCF and activates transcription. In the absence of an extracellular signal (called Wnt; see Chapter 3, Detecting Physiological Signals), β-catenin is continually destroyed by ubiquitin-mediated proteolysis. As we discussed for this form of degradation, phosphorylation of the substrate triggers ubiquitylation. Inactivation of a component required for this phosphorylation, or deletion of the site phosphorylated, leads to hyperactivity of β-catenin.

- Thanatropic dysplasia type II dwarfism is caused by a mutation in the activating loop of the kinase associated with the FGF receptor 3. This hyper-kinase activity produces an excess of activated STAT1.
- A single-amino-acid change in the so-called Eyk receptor converts a weak STAT3 docking site into a strong one. The consequence is inappropriate activation of STAT3 and tumorigenesis.

Interpreting Experiments

The nature of the regulatory systems we have described, once understood, alerts us to problems in interpreting experiments. Here are some considerations.

- Although anecdotal, it is widely recognized that overproducing an active kinase, whether in extracts or in cells in culture, can trigger activation of various signal transduction pathways. Sometimes the results of such experiments are taken to mean that there is a natural "cross-talk" between a kinase in one pathway and targets in another. But the result might also be an artifact (albeit an informative one) of overproducing the kinase; that is, the kinase, by virtue of being overproduced, might be working on substrates it ordinarily never sees, or ordinarily sees only weakly.

  A case in point is the Abl kinase. When separated from its inhibitory domains and overexpressed in cells, it activates signal transduction pathways promiscuously. This effect, incidently, is believed to explain how a truncated form of Abl (produced by a translocation) causes the human cancer chronic myelogenous leukemia.
- Even when an appropriate pathway is activated in such an experiment, the manipulation does not necessarily show how that pathway ordinarily works. For example, we have noted that overproducing a JAK and a STAT activates the STAT even in the absence of the receptor that is normally required to bring them together. Such "bypass" experiments might

be performed inadvertently: experiments with cell extracts or with cells in culture often include proteins present at unusually high levels.

- Identifying physiologically relevant protein-binding sites on DNA is difficult. Quite weak sites can be important if (as is usually the case) the specificity and affinity of binding of the protein are influenced by interactions with other DNA-binding proteins. Even where specific sites are clearly recognizable—for example, with λ repressor—small differences in affinities of those sites can be crucial to how the system works. Small differences in affinity may be measurable in vitro, but an assessment of their biological relevance requires more information than can be gathered by any one experimental approach.

  Identifying natural substrates for a kinase by examining protein sequences can also be problematic. The peptide sequences that can be phosphorylated occur in more proteins than the enzyme normally works on. A better match is found if one looks for the presence, on the protein, of two sites: one recognized by the active site of the kinase and another recognized by the docking module on that kinase.

- The results of certain "knockout" experiments might be misleading. The enzymatic function performed by the deleted protein can sometimes be provided by another enzyme that ordinarily does not work on the relevant substrate. A case in point is the result observed when a yeast MAP kinase was deleted. The substrate of the deleted kinase interacted with another MAP kinase, thus creating a kind of "cross-talk" between pathways. Presumably, this is ordinarily avoided because the substrate is sequestered by interaction with the "proper" MAP kinase.

Benefits

In the Introduction, we touched upon the idea that regulated recruitment (and its various modifications) has been extensively exploited by natural selection. The following are some suggestions for why this has happened.

- The same enzymatic active site can be used to produce different biological effects. For example, the same kinase active site can be directed to work on different substrates with vastly different biological consequences. The surmise that there are few stereospecific constraints on the relative dispositions of recruiting patches and the active sites suggests that these new specificities are easily evolved.

Analysis of the human genome sequence is just beginning as we write. Preliminary reports suggest that many of the "new" enzymes found in humans (i.e., enzymes not found in flies and worms) are actually old enzymatic active sites attached to new or additional recruiting domains.

- The meaning of signals is readily changed and expanded. For example, adding new sites for tyrosine phosphorylation to a receptor can expand the repertoire of enzymes and/or substrates brought to that receptor upon ligand binding. Analogously, adding binding sites for activators (or repressors) in front of a gene increases the number of signals that can switch on (or off) the gene. In both of these examples, there are evidently few geometric constraints, and the added sequences are short: about 4 amino acids for a new phosphorylation (on tyrosine) site and approximately 10 bp for a single regulatory protein-binding site in DNA.[13]
- It is conceivable that the use of similar enzymatic activities for different purposes allows a kind of redundancy that diminishes the negative effects of certain mutations. For example, perhaps the loss of one kinase can be compensated for, to some extent at least, by other kinases. Thus the damaged system would work, albeit with a reduced efficiency. To what extent this effect might hold is not known.

Our analysis of transcription in particular illustrated another reason why regulated recruitment is so readily exploited: there are many alternative ways a given regulatory problem can be solved using alternative combinations of identical or closely related regulators. Put another way, evolution is subject to fewer constraints than one might have expected from examining any one solution, the regulation of any one gene. Here are two illustrations of this matter, one from bacteriophage λ and the other from *Drosophila*.

Bacteriophage λ

In the λ switch, we described a mechanism that seems to depend on precise details: spacings between DNA sites, affinities of proteins for these sites and for each other, etc. But other phage, related to λ, employ closely related repressor and Cro proteins that interact with DNA sites arranged differently from their λ counterparts. Nevertheless, these switches work highly efficiently.

In each case, the regulatory proteins perform the same function as in λ. For example, repressor activates transcription of its own gene as it represses transcription of lytic genes. But, because activation can be effected by any of an array of contacts with polymerase, and because repression can be effected by any of an array of dispositions of repressor on DNA, these functions can be performed with various arrangements of sites.

We do not know, of course, whether these different phage switches have evolved independently or are the result of divergent evolution. But in any case, experimental manipulations indicate that there are hypothetical intermediate states in the evolution of the λ switch that work, but do so with decreased efficiency. Thus, for example, changing the affinities of various sites in the right operator for repressor produces phage that can grow lytically and form lysogens, but do so significantly less efficiently than the wild-type form.[14]

Drosophila

The stripe 2 enhancers found in different species of *Drosophila* are of similar size (750–950 bp) and bind the same regulators. But the precise arrangements and affinities of the regulatory protein-binding sites in the enhancers are not the same. Despite these differences, an enhancer from one species directs the proper pattern of gene expression (i.e., in stripe 2) when introduced into another species. Thus, there is more than one arrangement of binding sites that can bind the appropriate activators (and exclude the repressors) required to activate the *eve* gene in that region of the embryo.

Further study of the *eve* stripe 2 enhancer shows how readily a new response can be generated by reassorting pre-existing elements. Hybrid enhancers, containing the first half of the *eve* stripe 2 enhancer from one species fused to the second half from another, work, but they give slightly different patterns of *eve* expression. In one case, a sharp stripe of *eve* is produced, but its position has shifted slightly, and in the other case the stripe has broadened.

Certain distant relatives of *Drosophila* express *eve* in stripes, including one roughly corresponding to stripe 2. But the exact pattern of that stripe is different from what is found in *Drosophila* and that difference is believed to have important morphological consequences. Analyses of the natural and artificial *Drosophila* stripe 2 enhancers (as briefly reviewed here) suggest that nature can readily throw up functional variants for selection to work on.

FOOTNOTES

[1]The effector (i.e., the molecule that induces a conformational change) can be a small molecule such as a metabolite, but in other cases, the effector is a large molecule—another protein or even DNA, for example. In still other cases, the effector is an enzyme that covalently modifies the target protein—a kinase, for example. The term allostery is sometimes used in a more restricted context, referring to the trapping of a macromolecule in one of two alternate states.

In the Introduction, we emphasized how allostery can be used to control the level of activity of a protein (how well it works) but not its specificity (what it works on). There are exceptions to this rule. Recall from Chapter 3 (Footnote 1) the case of the glucocorticoid receptor, which is believed to interact with one or another of two protein complexes, depending on the site to which it is bound. In this case, it is suggested that the DNA site itself is an allosteric "effector," inducing one or another conformation to the bound receptor. But as this example illustrates, such effects are likely to be restricted to the choice between only a very limited set of alternatives, probably only two.

So allostery and recruitment both involve residues on an enzyme (for example) separate from its active site, but those residues serve quite different functions in the two forms of regulation.

[2]Not all enzymes that work on macromolecules are directed to substrate by recruitment as we use the term. For example, type II host restriction endonucleases specifically recognize short defined sequences in DNA, and that recognition is effected by amino acids in and around the enzymes' active sites. In that case, each active site is unique. The serine proteases use residues outside their common active sites to recognize substrates. But this example differs from those we emphasize in the text in that, rather than using auxiliary domains to effect this recognition, each protease has evolved a highly specific interaction with its unique substrate that involves broad surfaces of both protease and substrate. In that case, the active site bears a highly stereospecific relationship to the other residues that help bind substrate. As we have seen for RNA polymerases, for example, there are many ways to fruitfully recruit the enzyme to one gene or another, and we will see a similar theme in many of the examples in this chapter.

An enzyme that works on a small molecule can be regulated by recruitment if both enzyme and substrate interact with a third large molecule. For example, the enzyme PI3 kinase works on a lipid, a small molecule, but that lipid is attached to the cell membrane. Recruitment of the enzyme to the membrane facilitates the reaction.

[3]An overexpressed F-box protein lacking its substrate binding site (and fused to β-galactosidase) inhibits ubiquitylation. This is the result expected if the truncated F-box protein titrates the F-box binding site on the enzyme. It has been suggested that native F-box proteins are expressed constitutively, but are rapidly degraded by interaction with the machinery unless they are bound to substrate. This mechanism would ensure that only those F-box proteins required at any given time would be present at significant levels, and hence an inhibitory effect analogous to squelching would be avoided.

[4]Both classes of kinases are found in all multicellular eukaryotes, but simpler eukaryotes (e.g., the yeast *Saccharomyces cerevisiae*), like *E. coli*, lack tyrosine kinases. The transcriptional activator NtrC is activated by a histidine-specific kinase.

[5]The appearance and disappearance of different cyclins is determined by regulated synthesis and degradation, the latter involving the ubiquitylation system discussed earlier in this chapter. Full activity of a Cdk/cyclin requires phosphorylation of the Cdk's activating loop. That phosphorylation, in turn, is effected by a kinase (CAK) that works in response to mitogenic signals.

[6]STAT dimerization is also mediated by an SH2 domain on each monomer that recognizes a phosphorylated tyrosine on the other.

[7]The artificial tethering experiments were performed by attaching to each chain a protein that binds a bifunctional organic compound. The effect of the artificial dimerizer has been called "induced proximity." This strategy does not work for all cases. For example, the insulin receptor is not activated in this fashion. In that case, it is believed that insulin is required to hold the two chains together in the proper orientation for the cross-phosphorylation to occur. Put another way, some artificial dimerizers might hold the two chains in relation to one another such that the reaction is prevented.

[8]Ras is a member of a large family of proteins, called small GTPases. Each such protein adopts one conformation when bound to GTP, and another conformation when bound to GDP. Typically, the GTP-bound form binds to and activates a target protein, in this case, Ras-GTP binds to and activates the kinase Raf. SOS is an example of a G-protein exchange factor: it binds to Ras and removes GDP (which is produced by Ras-mediated hydrolysis of the bound GTP), freeing Ras to bind GTP.

Ras, activated by the interaction with SOS, in turn recruits and activates Raf, the first kinase in the so-called MAP kinase triad. The three MAP kinases—MAPkkk (in this case Raf), MAPkk, and MAPk—are believed to be held together on a protein scaffold. This localization ensures that a specific MAP kinase (the final enzyme in the pathway) is activated. Other MAP kinase pathways, held on other scaffolds, produce activated MAP kinases that work on different transcription factors and hence regulate different genes. Because there is limited specificity to the kinases upstream of the MAP kinases, scaffolding is required to impose specificity.

[9]Another, or additional, way to relieve this intramolecular inhibition is to remove the phosphate from the tyrosine residue recognized by the SH2 domain. Full activity, finally, requires phosphorylation of the activating loop. The inhibitory phosphorylation (that which creates the internal SH2-binding site) is mediated by a separate kinase called Csk.

[10]One way in which our picture of a generic STAT receptor was oversimplified (Figure 4.5) is that, although not shown in the figure, one phosphorylated tyrosine on the receptor recruits a phosphatase.

[11]The short protein sequences that are common sites for phosphorylation appear at a much lower frequency in eukaryotic proteins than in bacterial proteins. Perhaps this selection has been driven by unhealthy consequences of otherwise unavoidable phosphorylation.

[12]Introducing a constitutively active STAT3 into cells transforms them (i.e., renders them tumorigenic). This was accomplished by replacing the normal phosphorylation-dependent dimerization domains of the STAT (see Footnote 6) with a stronger, constitutive and presumably irreversible, disulfide bond.

[13]There are various ways that one could imagine bringing a gene under control of a set of regulators. One way would be, by recombination or transposition, to introduce new enhancers near the gene. Another way might be to mutate DNA stepwise; DNA binding is a quantitative matter, and one or a few changes could create a weak site. Even a weak site can have strong effects depending on the context, e.g., whether other interacting proteins bind nearby, etc. The fact that regulators can work when positioned at any of many places upstream or downstream from a higher eukaryotic gene would facilitate this process.

Whatever the detailed mechanism, the famous case of vertebrate eye lens proteins seems to show how readily a gene can be brought under control of a new regulatory element. In this case, several totally unrelated proteins, each of which has some other function (e.g., as a metabolic enzyme), have the appropriate optical properties to function as lens proteins. Which one is actually used for this purpose can differ even between closely related species. What happened, in each case, is that one or another of the possible genes was brought under control of the appropriate regulator so that the protein is expressed in the eye as well as in the original location.

[14]Such changes include making sites O_{R1} and $O_{R3.}$ of equal affinity for repressor. These experiments demonstrated that although optimum efficincy requires the relative affinities of these sites to be as found in λ (and related phage), the system could nevertheless work surprisingly well when the affinities were altered. This unexpected "robustness" is important when thinking about how the system evolved—it makes the stepwise accretion of improvements rather easy to imagine despite the apparent sophistication of the final outcome. Thus, for example, a site need not arise in the correct place and be of optimum affinity in a single step; even the wrong affinity will represent an improvement, one that can be further improved on by subsequent changes that alter its affinity for repressor.

BIBLIOGRAPHY

Ubiquitylation

del Pozo J.C. and Estelle M. 2000. F-box proteins and protein degradation: An emerging theme in cellular regulation. *Plant Mol. Biol.* **44:** 123–128.

Deshaies R.J. 1999. SCF and Cullin/Ring H2-based ubiquitin ligases. *Annu. Rev. Cell Dev. Biol.* **15:** 435–467.

Jackson P.K., Eldridge A.G., Freed E., Furstenthal L., Hsu J.Y., Kaiser B.K., and Reimann J.D. 2000. The lore of the RINGs: Substrate recognition and catalysis by ubiquitin ligases. *Trends Cell Biol.* **10:** 429–439.

Patton E.E., Willems A.R., and Tyers M. 1998. Combinatorial control in ubiquitin-dependent proteolysis: Don't Skp the F-box hypothesis. *Trends Genet.* **14:** 236–243.

Splicing

Graveley B.R., Hertel K.J., and Maniatis T. 1999. SR proteins are 'locators' of the RNA splicing machinery. *Curr. Biol.* **9:** R6–7.

TAT

Garber M.E. and Jones K.A. 1999. HIV-1 Tat: Coping with negative elongation factors. *Curr. Opin. Immunol.* **11:** 460–465.

Price D.H. 2000. P-TEFb, a cyclin-dependent kinase controlling elongation by RNA polymerase II. *Mol. Cell. Biol.* **20:** 2629–2634.

STATS (and SMADS)

Bromberg J.F. 2001. Activation of STAT proteins and growth control. *Bioessays* **23:** 161–169.

Bromberg J. and Darnell J.E., Jr. 2000. The role of STATs in transcriptional control and their impact on cellular function. *Oncogene* **19:** 2468–2473.

Darnell J.E., Jr. 1997. STATs and gene regulation. *Science* **277:** 1630–1635.

Ihle J.N. 2001. The Stat family in cytokine signaling. *Curr. Opin. Cell Biol.* **13:** 211–217.

Massagué J. and Chen Y.G. 2000. Controlling TGF-β signaling. *Genes Dev.* **14:** 627–644.

Massagué J. and Wotton D. 2000. Transcriptional control by the TGF-β/Smad signaling system. *EMBO J.* **19:** 1745–1754.

Kinases and Phosphatases

Cheng L. and Karin M. 2001. Mammalian MAP kinase signaling cascades. *Nature* **410:** 37–40.

Cohen G.B., Ren R., and Baltimore D. 1995. Modular binding domains in signal transduction proteins. *Cell* **80:** 237–248.

Hubbard S.R. 1999. Src autoinhibition: Let us count the ways. *Nat. Struct. Biol.* **6:** 711–714.

Hunter T. 2000. Signaling—2000 and beyond. *Cell* **100:** 113–127.

Kuriyan J. and Cowburn D. 1997. Modular peptide recognition domains in eukaryotic signaling. *Annu. Rev. Biophys. Biomol. Struct.* **26:** 259–288.

Mayer B.J. 2001. SH3 domains: Complexity in moderation. *J. Cell Sci.* **114:** 1253–1263.

Pawson T. 1995. Protein modules and signalling networks. *Nature* **373:** 573–580.

Pawson T. and Nash P. 2000. Protein-protein interactions define specificity in signal transduction. *Genes Dev.* **14:** 1027–1047.

Pawson T. and Scott J.D. 1997. Signaling through scaffold, anchoring, and adaptor proteins. *Science* **278:** 2075–2080.

Sharrocks A.D., Yang S.H., and Galanis A. 2000. Docking domains and substrate-specificity determination for MAP kinases. *Trends Biochem. Sci.* **25:** 448–453.

Sicheri F. and Kuriyan J. 1997. Structures of Src-family tyrosine kinases. *Curr. Opin. Struct. Biol.* **7:** 777–785.

Tonks N.K. and Neel B.G. 2001. Combinatorial control of the specificity of protein tyrosine phosphatases. *Curr. Opin. Cell Biol.* **13:** 182–195.

Zhou S. and Cantley L.C. 1995. Recognition and specificity in protein tyrosine kinase mediated signaling. *Trends Biochem. Sci.* **20:** 470–475.

General + Other Systems

Austin D.J., Crabtree G.R., and Schreiber S.L. 1994. Proximity versus allostery: The role of regulated protein dimerization in biology. *Chem. Biol.* **1:** 131–136.

Lee M. and Goodburn S. 2001. Signalling from the cell surface to the nucleus. *Essays Biochem.* **37:** 71–85.

Noselli S. and Perrimon N. 2000. Signal transduction. Are there close encounters between signaling pathways? *Science* **290:** 68–69.

Ptashne M. and Gann A. 1998. Imposing specificity by localization: Mechanism and evolvability (erratum in *Curr. Biol.* [1998] **8:** R897). *Curr. Biol.* **8:** R812–R822.

Sachs A.B. and Buratowski S. 1997. Common themes in translational and transcriptional regulation. *Trends Biochem. Sci.* **22:** 189–192.

Afterword

The difficulty lies, not in the new ideas, but in escaping the old ones, which ramify, for those brought up as most of us have been, into every corner of our minds.

JOHN MAYNARD KEYNES

WE HAVE TOLD OUR STORY WITH FEW ALLUSIONS TO HISTORICAL BACKGROUND. With the exception of issues raised in the Introduction, we have made little attempt to describe the order with which discoveries were made and concepts introduced and established. Such an undertaking would have entailed writing a far longer book. And, because (for example) key discoveries sometimes were made in yeast before being confirmed with bacteria, we think that a historical approach would have made it more difficult to present our extended argument.

The reader might then ask—and from our informal experience often does ask—to what extent are the ideas presented here controversial? To what extent is the book simply a summary of accepted ideas, and to what extent have we imposed our views? If the ideas are not novel, at what point in recent history did they become established? And so on.

We find these questions interesting but not easy to answer. In our experience—again speaking anecdotally—we find that the response to these questions varies considerably with the reader. For example, one reader found the book a "good summary" of what we know; others were surprised that a few simple formulations can evidently be applied so widely; and still others (we suspect) object to an unwarranted simplification of complex matters. We speculate here on some of the factors that might influence these responses.

Many of us were introduced to molecular biology by learning about classical enzyzmes—enzymes that have unique substrates recognized by precise active sites. Such enzymes, we learned, were integrated little (and sometimes not so little) "machines," the complex structures of which allowed them to be controlled by allostery. One of the beautiful aspects of the idea of allostery is that there is, in principle, no restriction between the chemical identity of the signal and the nature of the activity affected.

Jacques Monod famously remarked "with allostery, anything is possible." Many of us, we suspect, came to assume, perhaps subconsciously, that allostery must lie at the heart of all biological regulation. Allostery appeals to our innate love of complexity: each allosteric response requires evolution of a protein that responds in an appropriate way to the allosteric signal, and one would expect no general rules. That is, each allosteric enzyme would have its own complicated, integrated structural features that would allow the proper conformational change to the signaling chemical.

Of course, and as we mentioned in the Introduction, the very scientists who introduced us to allostery also showed how the same enzyme—bacterial RNA polymerase—was employed at different genes. Nevertheless, we suspect that many of us tacitly assumed we lived in an "allosteric world," a world based on beautifully evolved sets of machines that were turned on and off much as a car engine is switched on and off by the turning of a key.

The language we have used for many years reflects this bias: we speak of gene "activation" even for those cases (the majority we have argued) where neither the polymerase nor the gene is activated in any traditional sense. Rather, the enzyme is merely apposed with the substrate (in this case a specific gene) by an "activator."

"Activation" is a term widely used to describe control of other enzymes as well, often where the term is similarly inappropriate. For example, the F-box proteins of one of the ubiquitylating systems are called "ligases" and they are said to "activate" ubiquitylation of substrate. But as we have seen, the F-box protein is merely a carrier, one that apposes enzyme with substrate; it "activates" neither enzyme nor substrate. The terminology becomes problematic in a different way with proteins such as cyclins: these molecules do literally activate their target enzymes, but they also serve to appose the enzyme with one or more specific substrates.

And in these cases, and contrary to what a classical education might have prepared us for, there are limited stereospecific constraints on how the enzyme is initially apposed with the substrate. Thus, as experiments with kinases and RNA polymerases have shown, there are many ways to fruitfully appose an enzyme with a given substrate, using binding surfaces well separated from the active sites on the enzyme and the sites to be modified (or transcribed) on the substrate.

Seeing the general picture we have described has been hampered by the natural separation of fields, each with its own, usually forbidding, terminology. The discovery of the roles of "locator domains" (SH2s, SH3s, etc.) in the field of signal transduction was not immediately related to the roles of DNA-binding domains studied in the field of gene regulation. The problem is not just that the terminology is in each case so baroque: the paradoxical fact is that the very simplicity of the general mechanism we have been describing—determining enzyme specificity by apposition with alternative substrates—often requires a seemingly bewildering array of auxiliary proteins that make any given system work efficiently and subject it to proper controls. It is not so easy for an outsider (and sometimes even more difficult for an insider) to see the woods (assuming they are there!) for the trees.

A particular example shows us another kind of problem. *A Genetic Switch* (the book written by one of us some years ago) described many aspects of the molecular interactions governing gene regulation in bacteriophage λ. And many of these interactions turn out to be examples of interactions observed throughout nature. The book argued that λ repressor activates transcription of its own gene by binding DNA and contacting RNA polymerase binding to an adjacent promoter. But beyond that it was silent: what exactly was the function of that protein-protein interaction? As we have described here at some length in Chapter 1, results of certain physicochemical experiments seemed to prevent a coherent view that would link gene activation by λ repressor with that by CAP, for example. Only some years later, as we have described, did the advent of activator bypass experiments (first performed in yeast and then applied to the cases of λ repressor and CAP), taken with other findings, indicate that simple binding interactions could suffice to explain the actions of both these transcriptional activators.

We tend to find younger scientists least surprised by what we have to say. If that is generally true, perhaps it is because they are accustomed to hearing about matters that would strike the older variety as strange. They are used to hearing about (and working with) the two-hybrid system, in which a simple binding interaction triggers gene expression. They are used to regarding proteins (particularly eukaryotic ones) not so much as highly integrated units, but as collections of domains; and they are familiar with the notion that those domains, often, can be rearranged or separated from their natural neighbors—even attached to new neighbors—without loss of function.

That so much of the specificity of regulation—and hence so much of development and evolutionary change—depends on simple binding interactions is (or we think should be) hard to swallow. It certainly is for us. We, and we suspect many others, had expected that the meanings of biological signals would have been, somehow, more solidly based. As we have explained in the earlier part of this chapter, the rather crudely based systems are poised to go awry, and many of the complexities we see seem to be add-ons to get those systems to work.

It is understandable that we describe individual cases as "beautiful" and "elegant" (cf. the λ switch). But unlike Creationists (who revel in such descriptions), we realize that these systems evolved, stepwise. And so it should hardly be surprising that underlying all the complexities are certain rather simple mechanisms that, by being reiterated and constantly added to, can produce living systems.

APPENDIX 1

More on Cooperativity

The term cooperativity is used in many different contexts. We can state a more or less general definition of the term, but quite different underlying mechanisms can be involved in different cases.

We say that two ligands (e.g., small molecules or proteins) bind cooperatively to a third molecule if the binding of one ligand increases the binding of the other. In some cases, a conformational change—an allosteric effect—lies at the heart of the matter; in other cases, no such change is required. To illustrate this point, we begin with a classical case of cooperativity that involves a conformational change in the protein hemoglobin.

The binding of oxygen to one site on hemoglobin causes the protein to undergo a conformational change (or traps the protein in a particular conformation) such that the affinity for oxygen at a second site is increased. In this case, the oxygen molecules do not interact directly, and cooperativity depends on the change in state of the protein. Quantitative analysis of the curve describing oxygen binding to hemoglobin as a function of oxygen concentration (i.e., a so-called Hill plot) reveals that four oxygen molecules bind cooperatively to a single molecule of hemoglobin.

In the cases with which we have been most concerned—the cooperative binding of proteins to DNA—such allosteric changes are not required. Instead, adhesive protein-protein and protein-DNA interactions can mediate the effects. We expand upon this and related matters in the following sections.

TWO PROTEINS BINDING TO SEPARATE SITES ON DNA

Consider the binding of protein B to its DNA site in the presence and absence of protein A. Protein A, we assume, efficiently binds its DNA site even in the absence of protein B. Assume also that the two sites are close to each other, and that A and B touch each other when both are bound to DNA.

The interaction of A with B increases the concentration of B near its site, with B's DNA-binding surface free to interact with that site. An additional effect of the interaction between A and B can be to help pay the entropic cost of orienting B (e.g., restraining B from tumbling). Each of these effects increases the likelihood that B will, at any given instant, be bound to DNA.

In the example chosen above, A was prebound, but that is not necessary; just as A helps B to bind so B helps A to bind, and so the reaction could involve three components that assemble together: A, B, and DNA. Speaking roughly, the greater the differences in the intrinsic affinities of two sites (i.e., the affinity measured when the sites are separated from each other), the greater the differences in the respective "helping effects." Thus, if A binds much more tightly than B, the helping effect of A on B will be much greater than vice versa.

If the sites for A and B on DNA are not adjacent, the same rules apply, but now the additional energetic and entropic costs of looping the DNA must be taken into account as well. Other proteins binding to sites between those for A and B can help pay that cost by themselves bending DNA. Such proteins that bend DNA can bind cooperatively with other proteins that touch each other, even if the latter do not touch the bending protein. Another way that a third protein can help A and B bind, but without itself necessarily binding DNA, is by simultaneously contacting A and B. In this case, A and B will bind cooperatively without touching each other directly.

In some cases, it is not just a pair of proteins but multiple proteins (or multiple copies of the same protein) that interact and bind cooperatively. In such cases, the higher the number of such units that bind cooperatively, the steeper the curve describing the binding as a function of the protein concentration. We encountered such a steep curve in describing the cooperative binding of λ repressors to two DNA sites. In that case, the repressor monomers must form dimers, which then must interact to form a DNA-tethered tetramer, and that reaction proceeds roughly as the fourth power of the monomer concentration. The reaction thus proceeds in an "all or none" fashion over a small concentration range.

Of course, any measured difference in vitro may or may not be relevant in vivo—these cooperative effects will be relevant only over certain ranges of protein concentration. We noted in the text that weak interactions between DNA-binding proteins can have large effects on site occupancy. This follows from the familiar exponential relationship between binding energy and dissociation constant: $\Delta G = -RT\ln K_D$. This is why only a few kilocalories of binding energy between cooperatively binding proteins has such a dramatic effect on site occupancy.

We have emphasized that all of the effects described in this section can be achieved with simple binding interactions, ones that do not involve conformational changes in any of the reactants. But that restriction does not *necessarily* hold: interactions between DNA-binding proteins and between those proteins and DNA can, and sometimes do, involve conformational changes in one or another of the reactants. That conformational change could be extensive, revealing for example an otherwise buried DNA-binding surface. At the other extreme, there could be a subtle stiffening of the proteins that would decrease internal entropy. Still another effect could be on the DNA itself, stiffening the helix over a few base pairs, for example, to foster tighter binding. Some of these effects, particularly the more subtle ones, cannot be excluded in any given case without garnering a great deal of information.

ONE PROTEIN BINDING TO TWO SITES ON DNA

In this case, we have a stable oligomer with two surfaces that can simultaneously bind two separate DNA sites. In general, the presence of the second site helps binding to the first, but the magnitude of that effect will depend on the intrinsic affinities of the two sites. If the affinity of one (A) is higher than that of the other (A′), then the presence of site A can significantly increase the occupancy of A′; the presence of the weaker site will help the stronger to a lesser extent. We understand how binding to the strong site can help simultaneous binding to the weak site in the same terms we used to describe the cooperative binding of separate proteins to two sites, namely, by locally increasing the protein concentration and by entropic effects.

There is a salient difference between the two general cases that we have discussed: cooperative binding of two separate proteins to two sites vs.

binding of a stable oligomer to two sites. Thus, whereas in the first case, a sigmoid curve describes the binding of the proteins to DNA as a function of their concentrations, in the second case, it does not. As a consequence, increasing the number of sites (e.g., from one to two) recognized by a single protein (as with Lac repressor) can stabilize binding, or create a looped structure, but it cannot create a switch-like effect as seen with the binding of λ repressor.

For both of these general cases, any function that brings the interactants closer together will increase cooperativity. For example, if a DNA site is constrained to one part of the cell, bringing another site to that compartment will facilitate the interaction.

TWO PROBLEMS, ONE SOLUTION

Cooperativity seems to be the rule, rather than the exception, in describing protein-DNA interactions in gene expression. That is, proteins rarely seem to bind DNA without either interacting with other DNA-binding proteins or directly with two or more DNA sites (the λ Cro protein is an exception). This cooperativity solves two problems, as we now outline. The first involves the physicochemical problem of imposing specificity on protein-DNA interactions. The second involves the biological problem of integrating signals.

Specificity of protein-DNA interactions

Cooperative binding to DNA is a way to impose specificity without risking a kinetic problem that might otherwise arise. In brief, proteins that recognize specific sites with very high affinities would typically encounter a kinetic obstacle to finding those sites. This obstacle arises from the seemingly unavoidable increase in the affinity of the protein for nonspecific sites as the affinity for the specific site is increased. In addition, the concentration of nonspecific sites (essentially the DNA concentration) in cells is so high that the protein would be bound to such sites so tightly (and for such long periods) that it would never find its specific site.

The mutant Lac repressor X186 illustrates the problem. In a behavior opposite to that of wild-type repressor, this mutant represses transcription of the *lac* genes only in the *presence* of lactose (or of an ordinarily inducing derivative of lactose). The explanation is that the mutant binds DNA so tightly that, in the absence of inducer, it is sequestered on non-operator DNA sites. The inducer weakens the binding of the mutant repressor to the point where it resembles that of wild-type repressor in the absence of inducer. This renders the mutant capable of freeing itself from nonspecific DNA and binding to the *lac* operator.

The general strategy, therefore, to ensure tight and specific binding is to use cooperativity.

Signal integration

As we hope is clear from the text, cooperativity is a powerful tool for creating sensitive switches (as in the λ case) and for activating genes (where recruitment is the mechanism). It is also a powerful tool for integrating physiological signals, an example of which we saw in our discussion of the human interferon-β gene. In this case, three activators bind cooperatively to sites in the enhancer, and so the gene is activated only if all three activators are present in their active forms. This, in turn, depends on the simultaneous activities of three signal transduction pathways.

APPENDIX 2

Topogenic Sequences

IN THE TEXT, WE HAVE USED THE TERM RECRUITMENT in a rather restricted sense to refer to apposition of an enzyme with a chosen substrate. We have emphasized the use of adhesive surface patches to determine specificity. "Topogenic sequences" are peptides that direct (recruit) proteins from one cellular compartment to another. There are similarities between the properties of these sequences and activating regions that work by recruitment. Domain swap and other experiments similar to those we have described for transcriptional activators reinforce the parallels. We concentrate here on sequences that direct movement in and out of the nucleus.

Proteins destined for import bear characteristic short sequences called nuclear localization sequences (NLSs), and those destined for export bear nuclear export sequences (NESs). These sequences bind to carrier proteins called karyopherins which move back and forth through nuclear pores (themselves formed from about 30 proteins). Karyopherins (of which there are at least 12) release the cargo bearing an NES when outside the nucleus, and they release the cargo bearing an NLS when inside the nucleus.

Like many transcriptional activating regions, members of one class of NLSs (which bind one karyopherin) bear hydrophobic and charged residues and are likely to be unstructured when free of their targets. The charged residues on these NLSs are basic.

When present at high concentration, free NLS peptides block import. This effect is presumably like the "squelching" phenomenon that we encountered in our description of transcriptional regulation. In that case, targets in the transcriptional machinery were titrated by excess activating regions; in this case, the binding sites on the karyopherin are saturated by excess NLS. Attaching an NLS to a protein not normally found in the nucleus produces a hybrid that concentrates in the nucleus.

The structure of an NLS in complex with a karyopherin shows how hydrophobic and charged (basic in this case) residues can be recognized in a peptide lacking defined secondary structure, a theme we encountered in our discussion of transcriptional activating regions. It is deduced that hydrophobic interactions provide the bulk of the binding energy, and the spacing of the few charged residues determines specificity. It is tempting to imagine that this structure provides a general model for how peptides bearing hydrophobic and charged residues, including certain transcriptional activating regions, interact with their targets.

The karyopherins face a special problem: they must release cargo bearing an NLS when inside the nucleus, and release cargo bearing an NES when outside. How they do this is not understood in detail, but it is believed that the karyopherins adopt different conformations in the two cellular compartments. These conformations are determined by interaction with the small GTPase called RAN. RAN itself is found in one conformation in the nucleus (where it is bound to GTP) and in another in the cytoplasm (where it is bound to GDP). It is the GTP-bound form of RAN that interacts with, and changes the conformation of, the karyopherin. The two states of RAN, in turn, are determined by two accessory proteins called the RAN guanosine nucleotide exchange factor (RAN GEF) and the RAN GTPase-activating proteins (RAN GAP). RAN GEF promotes binding of GTP to RAN; it is located in the nucleus. RAN GAP promotes conversion of RAN-bound GTP to GDP; it is located in the cytoplasm.

APPENDIX 3

All-or-none Effects and Levels of Gene Expression

THERE IS NO INHERENT PROPERTY OF GENES, nor of the transcriptional machinery, that dictates only one level of expression of a given gene when that gene is "on." There are many examples in the literature, dealing with genes in bacteria, yeast, and higher organisms, that illustrate this point. Below, we describe two examples, one taken from yeast and the other from *Drosophila*.

- Expression of the yeast *HIS3* gene at its normal chromosomal location has been measured under the following three conditions: in the absence of any activator, and when stimulated by, in one case, a weak activator, and in another, by a strong activator. In the first of these scenarios, a low level of His3 is detected, and in the subsequent configurations, a successively higher level of the enzyme is produced. Moreover, the effect cannot be explained by saying that higher levels of expression merely reflect a higher proportion of cells expressing the enzyme at some uniform level. Rather, each cell in the population expresses an increased level of the enzyme in the successive configurations.

  This conclusion follows from the nature of the assay. Cells expressing His3 are resistant to 3-amino triazole (3-AT), with higher levels of His3 expression rendering cells resistant to higher concentrations of 3-AT. In the experiment described in the text, cells resistant to one level of 3-AT are uniformally sensisitive to a higher level of 3-AT. A variety of other experiments assaying single cells using fluorescent reporters, in both yeast and mammalian cells, have also demonstrated graded but uniform levels of expression of various genes.

- The *yellow* gene of *Drosophila*, whose product determines body pigment, provides another instructive example. The gene is expressed cell autonomously and so one can readily discern different levels of expression in individual cells in living flies. Experimental conditions that vary the efficiency with which the *yellow* enhancer can work yield different levels of expression, and these levels are similar in all relevant cells.

These experiments were performed by introducing the insulator that binds the protein Su(HW) between the enhancer and the gene. The efficiency of action of the insulator was varied by using various mutants of the insulator and by varying the number of insulator-binding sites.

Although the mechanisms are not clearly understood, the signal transduction pathways that trigger gene expression can work in an all-or-none fashion. For example, depending on the growth conditions, addition of successively higher concentrations of galactose to growing yeast cells can elicit full activation over a very narrow range of concentration of galactose. Under other growth conditions, activation increases more or less proportionally to the level of galactose concentration over a wide range. And we have described how cooperative binding of activators—in bacteria (λ) and in human cells (at the human interferon-β gene)—can achieve essentially all-or-none effects.

To summarize, the rate of expression of a given gene depends on many factors, including the inherent strength of the promoter, the number of activators and the strengths of their activating regions, and so on—and that level can be changed by varying any of a number of these parameters. For many genes, the particular level of expression seen has presumably been selected for. Whatever that level may be in any given case, cells can often interpret signals to get to that level in an all-or-none fashion.

Index